Postharvest Physiology and Storage of Tropical and Subtropical Fruits

– Author –
Richa Jha

Wordpen Academics
Publishing Science That Matter

ISBN: 978-93-90438-35-8 (Hardback)

Published by : **WORDPEN ACADEMICS**
4736/23, Ansari Road, Darya Ganj,
New Delhi – 110 002
Phone: +91-011-35004336
E-mail: info@wordpenacademics.com

Printed at **:** **Replika Press Pvt. Ltd.**

Exclusive Distributor: ***Bio-Green Books, New Delhi***

Preface

Post harvest losses of fruit a and vegetables are more serious in developing countries then those in well developed countries. An additional constraint to improving this situation is that in moat developing countries the number of scientists concerned with post-harvest food losses is significantly lower than those involved in production research. In the early days of horticulture in wolf developed countries, heavy losses occurred in much the same manner as they do today in developing countries. Increasing industrialization in technologically advanced nations gradually brought about improvements in crop handling. Elaborate harvesting equipment replaced the crude harvesting tools. Collection centres were strategically established in major producing areas. Containers were remodelled to add more protection to the produce. Commercial storage plants were installed and grade standards adopted. Engineers and economists became more and more aware of raw material behaviour. Concomitant advances in Refrigeration Technology in the developed countries have made possible establishment of cold chains for the entire post-harvest and handling operations. At the institutional level post-harvest research was initiated. Pilot packing houses were installed, coupled with the development of intensive training programmes. The improvement of product quality and reduction in post-harvest losses became the main concern of producers, middlemen, marketing specialists and consumers. Today, enormous volumes of quality horticultural crops produced in technologically advanced countries are made available to millions of people through improved post-harvest handling. Thus, historically and by necessity, post-harvest technology is part of the normal development processes in agriculture.

These handling procedures are not fully recognized in less developed countries. Here agriculture may be characterized as disjointed. Production is not linked with marketing. With perishable crops like fruits and vegetables, storage, packaging, transport and handling technologies are practically non-oxiatant, Hence, considerable amount a of, produce are loaf. Thus, as more fresh fruits and vegetables are needed to supply the growing population in less developed countries, as more produce is transported to non-producing areas, and as more commodities are stored longer to obtain a year-round supply, post-harvest lose prevention technology measures become paramount. It is distressing to note that so much time is being devoted to the culture of the plant, so much money spent on irrigation, fertilization and crop protection measures only to be wasted about a week after harvest. It is, therefore, important that post-harvest procedures be given as much attention as production practices the stages from planting until the product a roach the consuming public must be a mutual undertaking between the growers and those who will handle the products after harvest.

In the warm humid tropics, fresh fruit a and vegetables have an extremely, low level of natural protection against the climate peat a and bio-chemical and physiological deterioration. It is also in the warm humid tropics that human diseases are moat numerous and widespread and consequently where the need for the nutritional value of fresh fruits and vegetables la moat essential. On the other hand, dried grains and legumes, and to a certain extent, root crops (yams), have a fair degree of natural protection against the various deteriorative elements, even though considerable post-harvest losses do occur.

This is the first book to deal with the postharvest storage, physiology and conservation of all of the economically important tropical and subtropical fruits. The book is essential reading for all horticultural researchers and students working with these crops and for growers, exporters and importers within the industries concerned with tropical and subtropical fruits.

– Richa Jha

Contents

1

Introduction

The term fruit has different meanings dependent on context, and the term is not synonymous in food preparation and biology. Fruits are the means by which flowering plants disseminate seeds, and the presence of seeds indicates that a structure is most likely a fruit, though not all seeds come from fruits.

No single terminology really fits the enormous variety that is found among plant fruits. The term 'false fruit' (pseudocarp, accessory fruit) is sometimes applied to a fruit like the fig or to a plant structure that resembles a fruit but is not derived from a flower or flowers. Some gymnosperms, such as yew, have fleshy arils that resemble fruits and some junipers have *berrylike,* fleshy cones. The term "fruit" has also been inaccurately applied to the seed-containing female cones of many conifers.

Many true fruits, in a botanical sense, are treated as vegetables in cooking and food preparation because they are not sweet. These culinary vegetables include cucurbits, tomatoes, peas, beans, corn, eggplant, and sweet pepper; some spices, such as allspice and chilies, are botanical fruits. Occasionally, though rarely, a culinary "fruit" is not a true fruit in the botanical sense.

For example, rhubarb is often referred to as a fruit, because it is used to make sweet desserts such as pies, though only the petiole of the rhubarb plant is edible. In the culinary sense, a fruit is usually any sweet tasting plant product associated with seed(s), a vegetable is any savoury or less sweet plant product, and a nut is any hard, oily, and shelled plant product.

Technically, a cereal grain is a fruit termed a caryopsis. However, the fruit wall is very thin and fused to the seed coat so almost all of the edible grain is actually a seed. Therefore, cereal grains, such as corn, wheat and rice are better considered edible seeds, although some references list them as fruits. Edible gymnosperm seeds are often misleadingly given fruit names, e.g. pine nuts, ginkgo nuts, and juniper berries.

Fruit Development

The development sequence of a typical drupe, the nectarine (*Prunus persica*) over a 7½ month period, from bud formation in early winter to fruit ripening in midsummer.

A fruit is a ripened ovary. Inside the ovary is one or more ovules where the megagametophyte contains the mega gamete or egg cell. The ovules are fertilized in a process that starts with pollination, which involves the movement of pollen from the stamens to the stigma of flowers. After pollination, a tube grows from the pollen through the stigma into the ovary to the ovule and sperm are transferred from the pollen to the ovule, within the ovule the sperm unites with the egg, forming a diploid zygote.

Fertilization in flowering plants involves both plasmogamy, the fusing of the sperm and egg protoplasm and karyogamy, the union of the sperm and egg nucleus. When the sperm enters the nucleus of the ovule and joins with the megagamete and the endosperm mother cell, the fertilization process is completed. As the developing seeds mature, the ovary begins to ripen.

The ovules develop into seeds and the ovary wall, the *pericarp*, may become fleshy (as in berries or drupes), or form a hard outer covering (as in nuts). In some cases, the sepals, petals and/or stamens and style of the flower fall off. Fruit development continues until the seeds have matured. In some multiseeded fruits, the extent to which the flesh develops is proportional to the number of fertilized ovules.

The wall of the fruit, developed from the ovary wall of the flower, is called the *pericarp*. The *pericarp* is often differentiated into two or three distinct layers called the *exocarp* (outer layer, also called epicarp), *mesocarp* (middle layer), and *endocarp* (inner layer). In some fruits, especially simple fruits derived from an inferior ovary, other parts of the flower (such as the floral tube, including the petals, sepals, and stamens), fuse with the ovary and ripen with it. The plant hormone ethylene causes ripening. When such other floral parts are a significant part of the fruit, it is called an *accessory fruit*. Since other parts of the flower may contribute to the structure of the fruit, it is important to study flower structure to understand how a particular fruit forms.

Fruits are so diverse that it is difficult to devise a classification scheme that includes all known fruits. Many common terms for seeds and fruit are incorrectly applied, a fact that complicates understanding of the terminology. Seeds are ripened ovules; fruits are the ripened ovaries or carpels that contain the seeds. To these two basic definitions can be added the clarification that in botanical terminology, a nut is not a type of fruit and not another term for seed, on the contrary to common terminology.

There are three basic types of fruits:

1. Simple fruit

2. Aggregate fruit
3. Multiple fruit.

Simple Fruit

Epigynous berries are simple fleshy fruit. From top right: cranberries, lingonberries, blueberries red huckleberries.

Simple fruits can be either dry or fleshy, and result from the ripening of a simple or compound ovary with only one pistil. Dry fruits may be either dehiscent (opening to discharge seeds), or indehiscent (not opening to discharge seeds). Types of dry, simple fruits, with examples of each, are:

- achene – (dandelion seeds, strawberry seeds)
- capsule – (Brazil nut)
- caryopsis – (wheat)
- fibrous drupe – (coconut, walnut)
- follicle – (milkweed, magnolia)
- legume – (pea, bean, peanut)
- loment
- nut – (hazelnut, beech, oak acorn)
- samara – (elm, ash, maple key)
- schizocarp – (carrot seed)
- silique – (radish seed)
- silicle – (shepherd's purse)
- utricle – (beet).

Fruits in which part or all of the *pericarp* (fruit wall) is fleshy at maturity are *simple fleshy fruits*. Types of fleshy, simple fruits (with examples) are:

- berry – (redcurrant, gooseberry, tomato, avocado)
- stone fruit or drupe (plum, cherry, peach, apricot, olive)
- false berry – Epigynous accessory fruits (banana, cranberry, strawberry (edible part).)
- pome – accessory fruits (apple, pear, rosehip, saskatoon berry).

Aggregate Fruit

An aggregate fruit, or *etaerio*, develops from a flower with numerous simple pistils. An example is the raspberry, whose simple fruits are termed *drupelets* because each is like a small drupe attached to the receptacle. In some bramble fruits (such as blackberry) the receptacle is elongated and part of the ripe fruit, making the blackberry an *aggregate-accessory* fruit.

The strawberry is also an aggregate-accessory fruit, only one in which the seeds are contained in achenes. In all these examples, the fruit develops from a single flower with numerous pistils. Some kinds of aggregate fruits are called berries, yet in the botanical sense they are not.

Multiple Fruit

A multiple fruit is one formed from a cluster of flowers (called an *inflorescence*). Each flower produces a fruit, but these mature into a single mass. Examples are the pineapple, edible fig, mulberry, osage-orange, and breadfruit.

In the photograph on the right, stages of flowering and fruit development in the noni or Indian mulberry (*Morinda citrifolia*) can be observed on a single branch. First an inflorescence of white flowers called a head is produced. After fertilization, each flower develops into a drupe, and as the drupes expand, they become *connate* (merge) into a *multiple fleshy fruit* called a *syncarpet.*

There are also many dry multiple fruits, e.g.

- Tuliptree, multiple of samaras.
- Sweet gum, multiple of capsules.
- Sycamore and teasel, multiple of achenes.
- Magnolia, multiple of follicles.

Fruit Chart

To summarize common types of fruit:

- Berry – simple fruit and seeds created from a single ovary
  - o Pepo – Berries where the skin is hardened, like cucurbits
  - o Hesperidium – Berries with a rind, like most citrus fruit
- Epigynous berries(false berries) – Epigynous fruit made from a part of the plant other than a single ovary
- Compound fruit, which includes:
  - o Aggregate fruit – multiple fruits with seeds from different ovaries of a single flower
  - o Multiple fruit – fruits of separate flowers, packed closely together
- Other accessory fruit – where the edible part is not generated by the ovary.

Types of Fruit

Seedless Fruits

Seedlessness is an important feature of some fruits of commerce. Commercial cultivars of bananas and pineapples are examples of seedless fruits. Some cultivars of citrus fruits (especially navel oranges), satsumas, mandarin oranges table grapes, grapefruit, and watermelons are valued for their seedlessness.

In some species, seedlessness is the result of *parthenocarpy*, where fruits set without fertilization. Parthenocarpic fruit set may or may not require pollination. Most seedless citrus fruits require a pollination stimulus; bananas and pineapples do not. Seedlessness in table grapes results from the abortion of the embryonic plant that is produced by fertilization, a phenomenon known as *stenospermocarpy* which requires normal pollination and fertilization.

Seed Dissemination

Variations in fruit structures largely depend on the mode of dispersal of the seeds they contain. This dispersal can be achieved by animals, wind, water, or explosive dehiscence.

Some fruits have coats covered with spikes or hooked burrs, either to prevent themselves from being eaten by animals or to stick to the hairs, feathers or legs of animals, using them as dispersal agents. Examples include cocklebur and unicorn plant.

The sweet flesh of many fruits is "deliberately" appealing to animals, so that the seeds held within are eaten and "unwittingly" carried away and deposited at a distance from the parent. Likewise, the nutritious, oily kernels of nuts are appealing to rodents (such as squirrels) who hoard them in the soil in order to avoid starving during the winter, thus giving those seeds that remain uneaten the chance to germinate and grow into a new plant away from their parent.

Other fruits are elongated and flattened out naturally and so become thin, like wings or helicopter blades, e.g. maple, tuliptree and elm. This is an evolutionary mechanism to increase dispersal distance away from the parent via wind. Other wind-dispersed fruit have tiny *parachutes*, e.g. dandelion and salsify.

Coconut fruits can float thousands of miles in the ocean to spread seeds. Some other fruits that can disperse via water are nipa palm and screw pine.

Some fruits fling seeds substantial distances (up to 100 m in sandbox tree) via explosive dehiscence or other mechanisms, e.g. impatiens and squirting cucumber.

Uses

Many hundreds of fruits, including fleshy fruits like apple, peach, pear, kiwifruit, watermelon and mango are commercially valuable as human food, eaten both fresh and as jams, marmalade and other preserves. Fruits are also in manufactured foods like cookies, muffins, yoghurt, ice cream, cakes, and many more. Many fruits are used to make beverages, such as fruit juices (orange juice, apple juice, grape juice, etc) or alcoholic beverages, such as wine or brandy. Apples are often used to make vinegar. Fruits are also used for gift giving, Fruit Basket and Fruit Bouquet are some common forms of fruit gifts. Many vegetables are botanical fruits, including tomato, bell pepper, eggplant, okra, squash, pumpkin, green bean, cucumber and zucchini. Olive fruit is pressed for olive oil. Spices like vanilla, paprika, allspice and black pepper are derived from berries.

Nutritional Value

Fruits are generally high in fibre, water and vitamin C. Fruits also contain various phytochemicals that do not yet have an RDA/RDI listing under most nutritional factsheets, and which research indicates are required for proper long-term cellular health and disease prevention. Regular consumption of fruit is associated with reduced risks of cancer, cardiovascular disease, stroke, Alzheimer disease, cataracts, and some of the functional declines associated with aging.

Nonfood Uses

Because fruits have been such a major part of the human diet, different cultures have developed many different uses for various fruits that they do not depend on as being edible. Many dry fruits are used as decorations or in dried flower arrangements, such as unicorn plant, lotus, wheat, annual honesty and milkweed. Ornamental trees and shrubs are often cultivated for their colourful fruits, including holly, pyracantha, viburnum, skimmia, beautyberry and cotoneaster.

Fruits of opium poppy are the source of opium which contains the drugs morphine and codeine, as well as the biologically inactive chemical theabaine from which the drug oxycodone is synthysized. Osage orange fruits are used to repel cockroaches. Bayberry fruits provide a wax often used to make candles. Many fruits provide natural dyes, e.g. walnut, sumac, cherry and mulberry. Dried gourds are used as decorations, water jugs, bird houses, musical instruments, cups and dishes. Pumpkins are carved into Jack-o'-lanterns for Halloween. The spiny fruit of burdock or cocklebur were the inspiration for the invention of Velcro.

Coir is a fibre from the fruit of coconut that is used for doormats, brushes, mattresses, floortiles, sacking, insulation and as a growing medium for container plants. The shell of the coconut fruit is used to make souvenir heads, cups, bowls, musical instruments and bird houses.

Fruit is often also used as a subject of still life paintings.

Ethylene in Fruit Physiology

The apple is a remarkable fruit. When harvested at optimum storage potential, it may store for as long as 12 months under proper temperature and atmospheric conditions. Correct harvest timing is based on defining the stage of maturity of the apple, which may be best determined by measuring fruit ethylene biosynthesis. Ethylene was discovered in the late 1800's. Like many other discoveries, it was somewhat accidental. Someone had observed several trees along a tree-lined boulevard had lost their leaves. It was soon discovered that there was a leak in one of the underground pipes carrying natural gas. It didn't take long to learn ethylene was the causative agent. Since then, hundreds of papers have been written on ethylene action, biosynthesis, promotion, inhibition, and molecular genetics.

Ethylene is found in most living tissues. In most terrestrial mammals, small amounts of ethylene are expressed with every exhaled breath, but in animals, ethylene is not considered a hormone as it is in plants. There are seven major effects of ethylene in plants which are: promoting ripening, inducing fruit abscission, inducing flowering, promoting seed germination, breaking dormancy, promoting root initiation and inducing vegetative dwarfing. In the science of fruit growing, most of these effects are put to use for the grower's benefit. This paper gives an overview of ethylene's involvement in the process of fruit ripening, how it is measured, and what the measurements mean.

Fruit Development

Ethylene production is largely pre-determined in both time and amount by the genetics of the fruit and, depending on when the fruit blossoms and the climate, will determine to a large degree the development of the respiratory climacteric. During fruit development, respiration (i.e., the generation of carbon dioxide), which is a measure of metabolic activity, declines gradually throughout the season until several weeks before it ripens where it reaches what is known as a preclimacteric minimum. At this point, the metabolic functions of the fruit are in a near resting stage in preparation for a burst of metabolic activity signifying ripening. During ripening both carbon dioxide and ethylene increase significantly. The main developmental stage of the fruit is referred to as maturation, in which photosynthate is converted to starch. The ripening phase is when the starch is converted to sugar. Senescence is that stage in which the membrane functions break down due to degradation of lipid bilayers leading to cell damage and necrosis.

Optimum harvest is a subjective measurement defined as a fruit with good keeping quality and good eating quality. If picked too early during the maturation stage, insufficient starch will be converted to sugar and the fruit will keep well enough, but the eating quality will be poor. On the other hand, if the fruit is picked too late, there will be insufficient starch and acid reserves for metabolic maintenance in storage, but the eating quality will be good. Therefore, the optimum timing is critical to the proper storage (and marketing) of apples. Researchers throughout the world have spent decades defining the optimum time of harvest for long-term CA storage and yet, as new varieties enter the marketplace, optimums must be redefined for each individual cultivar.

Ethylene Production

After the respiratory preclimacteric minimum comes the respiratory climacteric. This is the point at which the fruit will generate a high, sustained respiration and ethylene biosynthesis. It occurs at different times for different cultivars. For example, a Yellow Transparent apple, which ripens in early July, has a very quick respiratory climacteric and may ripen within a few days. In fact, there are times when I have gone out one day and seen green fruit on the tree and two days later, these fruit are yellow and on the orchard floor.

The respiratory climacteric of this apple is quick and peak is high. On the other hand, a Delicious apple, which ripens sometime in mid-September, has a somewhat longer climacteric period. This fruit may develop a respiratory climacteric over a period of 10 days, for example. This is advantageous to a grower because it allows time to harvest all of the fruit without being so concerned that the fruit will ripen early and even drop from the tree. Further, there are cultivars that ripen later, such as Fuji, Braeburn, or Granny Smith. These cultivars in turn have an even broader climacteric and the peak of respiration (when similarly measured) is lower than in the previous scenarios.

Consider, for example, ethylene from the cultivar Braeburn. In one of our trials, fruit was picked over a period of 10 weeks beginning around August 15. If we measure the ethylene from the core by sampling with a syringe 24 hours after the fruit has been harvested, we see a very low ethylene production within the fruit tissue—generally less than 2 ppm. On the other hand, if we wait for four days before sampling the core gases of the fruit, we see significantly higher levels of ethylene reaching about 10 ppm at proper harvest. Waiting for seven days after the fruit is harvested before sampling, results in ethylene values in the 60-70 ppm range at harvest time. This suggests that delaying cooling (storage) of the fruit after it has been picked hastens the ripening process and, therefore, is deleterious to the condition of the fruit out of storage. This is important because often times a grower will harvest his fruit in bins and it may stay in his orchard or on the warehouse receiving dock before it is properly cooled. The longer the delay between harvest and storage the shorter will be the storage life and the poorer the condition of the fruit after it is removed from storage.

Sampling Fruit Ethylene

Clearly, the accurate measurement of ethylene in apples is vital to their proper storage and therefore their condition out of storage. But with so many fruit (100 million boxes of roughly 100 apples each in Washington State alone, and at least 4 times that throughout the world) the problem has become not the accuracy of sampling, but the unfathomable number of samples to assess.

Early efforts to analyse gas samples used a bulk approach. A bushel of apples would be placed in a large glass container and a sample of the headspace gas would be withdrawn and sampled. If each apple were very similar, this approach would accurately depict the physiological state of the individual fruit and, therefore, the proper regime for optimum storage. This is hardly the case. Data obtained from this method was qualitative, however, because there always exists the possibility that one apple could be riper than the others, thereby generating most all the ethylene. Thus, although all of the other fruit could have extremely low ethylene levels, the single mature apple would indicate the "average" ethylene was quite high. Because there are differences due to size, colour, tree position, stress, variety, rootstock, soil, and a host of other factors, the best that can be said of this method is that it would always overestimate the average ethylene production.

Individual apples were then evaluated for ethylene production. The most common method was to insert an 18-gauge needle into the apple through the calyx end extending through to the core. A 0.5 to 1.0 mL sample of gas could be withdrawn and subsequently analysed by gas chromatography. (Routine analysis of ethylene by megabore gas chromatography is sensitive to the ten parts per billion range.) By increasing the number of samples and examining the variability among samples, one could obtain a more accurate picture of the physiological state of the part of the orchard sampled.

Harvest decisions could then be based on both the amount of ethylene being produced as well as on the degree of variation of samples. High ethylene blocks (experimental sites within and orchard) with high variability might well indicate that very few fruit are producing significant amounts, whereas, high ethylene blocks with little variability might indicate the entire block is ripening more evenly. The concern with this method lies in not knowing the physiological meaning of a gas sample taken from the fruit centre. Again, there are differences due, for example, to size of seed cavity, amount of water in the fruit, amount of interstitial space, cell density and the amount of wounding from the needle insertion. Generally, any damage to the fruit tissues induces ethylene biosynthesis.

The method presently under design of assessing individual fruit ethylene is to place each apple in a small, 3-liter Plexiglas chamber with a low flow of scrubbed air. Micro-processors may sample the effluent gas as often as required 24 hours a day. This takes a single person about 1 hour to set up the system capable of analysing gas composition of hundreds of apples—a process that manually takes about 2 to 3 minutes per sample. The greater number of samples increases the likelihood that apples will reach the consumer with optimum quality.

What is needed is a method to assess quickly and accurately the ethylene generated by the whole apple, or one that uses part of the fruit that is most directly correlated with the physiological stage of fruit maturity. As for sampling a portion of tissue, it almost seems as though Heisenberg's principal is at work here in that any invasive action to the fruit will change its physiological stage and response.

Perhaps the closest thing the apple industry has to assessing individual fruit maturity is in the fruit sorting process. In modern fruit packing warehouses individual fruit are mechanically sorted into weight categories and optically scanned for colour sorting. Not far away are sensors for inline determination of firmness, sugar content and internal disorders. Practically everything can be determined about an individual fruit without damaging it. So far, however, only ethylene gives a meaningful picture of the physiological stage of ripeness. Thus, the problem remains; how does the apple industry nondestructively assess ethylene of 10 billion apples.

Inhibiting Ethylene Biosynthesis

Abbott Laboratories has recently developed a new chemical tool for use in apple production. AVG (aminoethoxyvinylglycine) was discovered in 1976 by researchers at Hoffmann-LaRoche, but it wasn't until the last few years that it was made commercially available. This compound is a substituted amino acid produced by a streptococcal mold and is specifically targeted to inhibit the enzymatic production of ACC, which is the precursor of ethylene. In 1980, Williams in Wenatchee showed that AVG could be used to inhibit internal ethylene production thereby reducing fruit abscission, but could not be used to compensate for external ethylene, no matter the source. In 1981, we did some experiments on Golden Delicious apples, in which we took fruit that was harvested at optimum harvest time and dipped them in solutions containing 0, 100, 200, or 400 ppm AVG (Curry, et al.). These fruit were stored in Plexiglas chambers and the ethylene and carbon dioxide were measured daily for a period of seven weeks. Ater about 10 days, the untreated fruit showed an increased rate of respiration that coincided with the climacteric rise. For this treatment, ethylene exceeded 50 ppm for the remainder of the study. Fruit dipped in 100 ppm AVG had a respiratory climacteric delayed about six weeks. At seven weeks, ethylene levels were about 20 ppm and by the ninth week, ethylene had not exceeded 30 ppm. At 200 ppm, it took approximately nine weeks to see a slight increase in carbon dioxide and at that time, ethylene was approximately 4 ppm. Lastly, at 400 ppm, Golden Delicious apples dipped after harvest never showed an increase in respiration and the ethylene levels were near zero, even at nine weeks after harvest. At the end of the nine weeks, the control apples were yellow and shriveled whereas apples dipped in AVG were still green and noticeably firmer.

In another trial, Williams (1980) treated Delicious apples with pre-harvest applications of AVG approximately three days before harvest. These fruit were stored in regular cold storage and after three weeks, showed no difference from those that were not treated with AVG. By 16 weeks, however, differences began to occur. Firmness of untreated apples was 13.9 pounds, whereas those treated with 450 ppm

AVG three days before harvest showed an average firmness of 15.2. Thirty weeks after harvest control fruit had a firmness of 12.8 pounds, whereas those treated with AVG were a full pound firmer.

Low Ethylene in Controlled Atmosphere

Controlled atmosphere was developed to provide for the storage of apples as production increased throughout the United States and the world. Controlled atmosphere that pertains to fruit storage is defined simply as the regulation and control of major gases, such as oxygen and CO_2, and minor gases, such as ethylene or other volatiles. This is in contrast to an uncontrolled atmosphere or regular storage where none of the gases are regulated, but may be monitored. This atmosphere also changes composition as the apples continue to respire, but it depends entirely upon the venting of the room.

Before controlled atmosphere storage, apple storage was limited to the time that the apple would remain marketable (firm) in regular cold storage. This might be either in refrigerated units or natural caves in countries where energy is limited and therefore expensive. With new technology, however, apples may be stored for as long as 12 months. This complicates a grower's life in that in the 12th month of storage, an apple producer has to be considering three apple crops. The first would be the crop still in storage after 12 months. The second would be the current crop that is now on the tree waiting to be harvested. The third would be that which is on the tree in the form of a bud, which has a primordial flower and which is capable of sustaining winter injury. That's a lot of apples to consider at any one time.

Physiology of Pruning Fruit Trees

Woody plants are pruned to maintain a desired size and shape and to promote a certain type of growth. Ornamental plants are pruned to improve the aesthetic quality of the plant, but fruit trees are pruned to improve fruit quality by encouraging an appropriate balance between vegetative (wood) and reproductive (fruiting) growth. Annual pruning of fruit trees always reduces yield, but enhances fruit quality. Pruning increases fruit size because excess flower buds are removed and pruning encourages the growth of new shoots with high-quality flower buds. Pruning improves light penetration into the canopy, and light is required for flower-bud development, fruit set and growth, and red colour development. Pruning also makes the canopy more open and improves pest control by allowing better spray penetration into the tree; air movement throughout the canopy is increased, which improves drying conditions and reduces severity of many diseases.

This publication describes why plants respond to pruning and other forms of plant manipulation used to train trees. This information applies to all plants, but application to fruit trees is emphasized.

Pruning fruit trees is somewhat of an art based on an understanding of plant physiology and development. In other words, if we understand how plants grow and how they will respond to different types of plant manipulations, we can alter vegetative growth and fruiting to obtain trees and fruit with desirable characteristics.

A basic understanding of certain aspects of plant physiology is a prerequisite to understanding pruning. Unlike animals, plants continue to increase in size throughout their lives. There are only two ways plants can grow.

Primary growth is the increase in length of shoots and roots, and is responsible for increases in canopy height and width.

Secondary growth is the increase in thickness of stems and roots.

Both types of growth require cell division followed by cell enlargement and differentiation.

Plant Growth

Meristems are regions of cell division and there are two types of plant meristems. An apical meristem is located at the tip of every shoot and root. As cells divide in these apical meristems, the shoots and roots elongate as cells are piled one on another. Behind the region of cell division is a region of cell differentiation, where cells enlarge and differentiate into various tissues. In the axil of each leaf is a small apical meristem called an axillary meristem that forms an axillary bud, which usually remains dormant until well after the subtending leaf is fully developed. An axillary bud may remain dormant or develop into a lateral branch or a flower.

There are two distinct layers of meristematic tissue within the stem or root responsible for secondary growth, the vascular cambium and the cork cambium. The vascular cambium is a cylinder of specialized cells, usually five to ten cells thick, running the length of the plant, including the roots, and is responsible for the radial growth of plant parts.

Phloem cells are produced to the outside of the cambium and xylem cells are produced to the inside of the cambium. Downward transport of sugars, nutrients, and hormones from the top of the tree to the roots occurs in the phloem tissue. Xylem cells are tube shaped, become hollow and die to form a pipe-like system through which water, hormones and mineral nutrients move from the roots to the top of the tree. Most of the radial growth of woody plants is due to activity of the vascular cambium, but a small amount results from activity in another lateral meristem, the cork cambium, located outside the vascular cambium.

The cork cambium together with the cork cells, constitute the periderm: a protective layer of suberized dead cork cells forming the bark. Suberization is the impregnation of cell walls of cork tissue with a fatty substance called suberin. Each season new layers of cells are produced and appear as growth rings when viewed in cross-section.

Over time, the xylem cells at the centre of the trunk or limb are crushed and become nonfunctional as transport pipes, but they do provide structural support to hold the plant upright. While grafting it is important to line up the cambiums of the scion and the rootstock to ensure a successful graft union.

Buds

Buds are important to the vegetative and reproductive growth of trees. Fruit tree training and, to a lesser extent, pruning primarily involves bud manipulation. Buds are actually undeveloped shoots. When a vegetative bud is sliced longitudinally during the winter and viewed under magnification, the apical meristem at the tip, leaf primordial (developing leaves), axillary meristems, developing axillary buds, and procambial tissue (tissue that will develop into the cambium) are all visible.

Buds on fruit trees usually have about seven leaves and initial shoot elongation in the spring results from cell expansion. During late June and July some of the shoot apices will flatten out and develop into flower buds. Flower buds are actually modified shoots and the various flower tissues are actually modified leaves. Although the process of switching from vegetative to reproductive buds is not fully understood, hormones that can be influenced by environmental factors, stresses, and plant nutrition control the process.

There are several things we can do to influence whether or not a bud becomes a flower bud or remains vegetative. In general, factors that favour rapid growth, such as high nitrogen levels in the shoot tissues, inhibit the development of flower buds. Applying growth-promoting plant growth regulators such as gibberellins usually inhibits flower-bud induction, whereas ethylene may promote flower-bud development. Mild stresses such as shoot bending and water stress may also promote flower-bud development.

Producing annual crops of high-quality fruit requires a balance between reproductive and vegetative growth. Fruit producers use various techniques, including pruning, branch bending, and plant growth-regulator sprays, to manipulate tree growth and flowering. Often these techniques affect bud dormancy, so knowledge of buds and bud dormancy is essential if we are to understand how pruning influences tree growth. It is also important to be able to identify the different types of buds on a tree, especially to distinguish between flower and vegetative buds.

Buds may be classified as to location, contents, or activity.

Classification by content Several types of buds commonly develop on fruit trees. Vegetative buds only develop into leafy vegetative shoots. Flower buds produce only flowers. Stone fruit trees (peach, nectarine, apricot, plum, and cherry) produce vegetative buds and flower buds. Apple and pear trees produce vegetative and mixed buds. Both leafy shoots and flowers emerge from mixed buds.

Classification by Location

Terminal buds are located at the tip of a shoot. On stone fruit trees terminal buds are vegetative buds. Terminal buds on apple and pear trees are usually vegetative; however, some varieties such as Rome Beauty produce mixed buds terminally and are referred to as tip bearers or terminal bearers. Most mixed buds on apple and pear trees are formed terminally on short, less than sixinch, shoots that terminate with a rosette of leaves.

These short shoots are called spurs. Lateral buds form in the axils of leaves and are often referred to as lateral buds or axillary buds. On stone-fruit trees, lateral buds may be either vegetative or flower. Nodes on one-year-old shoots may have one to three buds, some of which may be flower buds and others vegetative buds. Flower buds are larger with tips that are relatively round, whereas vegetative buds are small, narrow, and pointed. In the case of apple and pear trees, lateral buds on the previous season's growth are usually vegetative. However, lateral buds on some varieties, especially on the dwarfing rootstocks, may be mixed buds.

Classification by Arrangement on the Stem

The bud arrangement influences the arrangement of a fruit tree's branches and thus the tree's shape and how easy it is to manage. A node is the joint on a stem where a leaf is or was attached. Axillary buds are located in the axis above where a leaf is attached to the stem. In apples there is usually only one leaf per node, whereas three leaves often arise from a node on peach shoots. When a leaf falls in the autumn, a leaf scar remains just below the axillary bud. Buds are opposite when there are two at the same node but on opposite sides of the stem. Forsythia is an example of a plant with opposite buds. Buds are alternate when there is only one from each node and no one bud is on the same side of the stem as the one next above or below it. Deciduous fruit trees have buds that spiral along a shoot. The spiraling three-dimensional arrangement of leaves around a stem is known as Phyllotaxy and is expressed as a fraction, where the numerator is the number of turns to get to a leaf directly above another and the denominator is the number of buds passed.

Classification by Activity

Buds are dormant when they are not visibly growing. When shoots develop around large pruning cuts, they usually are sprouting from dormant buds. Adventitious buds form irregularly on older portions of a plant and not at the stem tips or in the leaf axils. They form on parts of the root or stem that have no connection to the apical meristems. They may originate from either deep or peripheral tissues.

For example, shoots often arise from adventitious buds growing from callus tissue around wounds. Root suckers (vigorous upright shoots developing from the roots) develop from adventitious buds on the roots.

Plant Hormones

Hormones are substances produced in very small amounts in one part of the plant and transported to another part where they cause a response. Plants produce a number of hormones that control various aspects of growth, such as stem elongation; dormancy of buds and seeds; flowering; fruit set, growth, and ripening; and the response to light and gravity. While pruning, it is useful to consider the activity of the general types of hormones, promoters (gibberellins and cytokinins) and inhibitors (auxins and abscisic acid). Promoters generally cause bud growth, cell division and elongation, and stem growth. Inhibitors are usually associated with dormancy and inhibit shoot development from seeds and buds and may be involved in flower-bud induction. It is often the ratio of promoters and inhibitors, rather than their absolute concentrations, that determines how a plant will grow. The production of plant hormones is usually controlled by environmental conditions such as temperature or day length. Vegetative growth is usually associated with low ratios of inhibitors to promoters and dormancy is usually associated with high ratios of inhibitors to promoters.

Dormancy is a condition characterized by temporary growth cessation and suppressed metabolism. During the winter trees appear not to be growing, but the tissues are alive, there is metabolic activity, and cells are slowly expanding and differentiating. By early October all the flower parts can be seen in a flower bud and vegetative buds contain leaves. During the winter these various tissues continue to enlarge and differentiate. Given favourable growth conditions, some buds will develop into shoots or flowers, but others may remain dormant. By understanding the factors influencing bud dormancy we often can influence certain aspects of tree growth. Buds of deciduous trees go through several stages of dormancy. Results from dormancy research are confusing because plant physiologists have used different terminology to describe the stages of dormancy. Plant physiologists currently describe dormancy in four stages.

Para-dormancy occurs in the mid to late summer when buds do not grow because inhibitors produced in the leaves and terminal buds inhibit bud growth. Para-dormancy can often be overcome by removing leaves (leaf stripping) along a section of a shoot so the axillary buds develop into shoots. Nurserymen often use this technique to produce trees with lateral branches (feathered trees). Using heading cuts to remove the terminal portion of a shoot will allow several axillary buds just below the cut to develop into shoots. Sometimes an application of growth promoters (gibberellins and/or cytokinins) will induce bud growth.

Sometimes axillary buds do not become dormant and develop into shoots within a few days of being formed. Such shoots are referred to as sylleptic shoots and are fairly common on vigorously growing peach trees, but are rarely produced on apple trees.

Ecto-dormancy occurs in the early fall, before defoliation, when plants do not grow because the environmental conditions are not conducive for growth. Growth will resume if the plants are exposed to suitable temperatures and day lengths.

Endo-dormancy occurs during the winter because there are high levels of inhibitors (abscisic acid) within the buds. During this phase of dormancy the trees will not grow even under ideal growing conditions. The concentration of inhibitors declines as buds are exposed to chilling temperatures. Temperatures near 45°F are ideal for chilling, but temperatures between 35° and 55° F will provide some chilling. The chilling requirement to satisfy dormancy for most varieties of apples and peaches grown in Virginia is about 1,000 and 800 hours, respectively. When the chilling requirement is satisfied, the level of inhibitors within the bud is low enough that growth may commence when environmental conditions are appropriate for growth. Avoid planting varieties with chilling requirements less than 800 hours because such varieties usually bloom early and are susceptible to frost.

Eco-dormancy occurs in the late winter, usually by mid January, after the chilling requirement has been satisfied. At this time the trees do not grow because conditions are unsuitable for growth. Growth will commence when trees are exposed to warm temperatures.

Apical dominance is a type of para-dormancy, where axillary bud growth is inhibited in the apical meristematic zone. Axillary buds on fruit trees typically remain dormant for a prolonged period while the main shoot continues to grow. Apical dominance has been studied for more than 80 years, and the exact mechanism is not yet fully understood, but it seems to be controlled by the relative concentrations of inhibitors and promoters. Growth of axillary buds is inhibited by high concentrations of auxin produced by the terminal bud. Auxin moves down the shoot, from cell to cell by gravity, so concentrations are highest near the shoot tip. Promoters are produced in the roots and are transported upward in the tree. Growth of axillary buds may occur at the base of shoots where concentrations of inhibitors are relatively low and concentrations of promoters are relatively high.

You can overcome apical dominance by removing the shoot tip, which is the source of auxin. The three or four buds immediately below a heading cut usually develop into shoots. Pinching annual plants to induce branching is a form of heading. Another way to overcome apical dominance is to notch buds. Notching involves cutting through the bark to hard wood, with a knife or hacksaw blade at about bloom time, just above a bud. The cut interrupts the downward flow of inhibitors, but not the upward flow of promoters, and releases the bud from dormancy. On vigorous upright one-year-old shoots, notching often successfully overcomes dormancy in about 70 percent of the buds. Sometimes apical dominance can be overcome by spraying shoots with promoters just before bloom time.

Shoot Bending

Shoots bend in response to an auxin gradient within the shoot. Everyone who has grown plants in the house has noticed that plants tend to grow towards the light. This phenomenon is known as photomorphism and is caused by varying concentrations of auxin in different sides of a stem or shoot. Auxin causes cells to elongate, but auxin is destroyed by light. Therefore, there is a higher concentration of auxin on the dark side of a shoot and the cells on the dark side elongate more than cells on the sunny side of the shoot, causing the shoot to bend towards the light.

Tree fruit producers have noticed a similar phenomenon where the tips of growing branches tend to bend upward, even when the branch was physically oriented to the horizontal. This condition, known as gravimorphism, is also caused by an auxin gradient within the branch in response to gravity. Auxin flows by gravity to the lower side of a limb. The subsequent accumulation of auxin is responsible for increased cell elongation on the underside of the limb, and the growing tip bends upward.

Another consequence of gravimorphism is the development of watersprouts from the upper surface of horizontally oriented limbs. Watersprouts are vertically growing shoots that develop from the upper surfaces of branches or near pruning cuts. High auxin concentrations on the underside of the limb inhibits growth of buds on the underside of the limb, but the concentration of auxin on the upper side of the limb is inadequate to inhibit bud growth and many of these buds develop into watersprouts. Watersprouts are usually undesirable and their development can be suppressed by orienting limbs no more than 45 degrees from the vertical. Fruit trees are sometimes trained as espalier. There are several ways to espalier trees, but one method involves orienting limbs to a horizontal position. This system induces many watersprouts along the length of the branches. Watersprout development can be greatly suppressed by orienting limbs 45 to 60 degrees above horizontal.

Reducing tree height by cutting into large diameter branches or trunks often results in the development of vigorous watersprouts around the cut. There are buds buried in the bark that normally remain dormant. However, a severe pruning cut will release these buds from dormancy.

Fruit Cultivation in India

The exhaustive and comprehensive cultivation of vegetables, flowers and fruits is referred to as horticulture. Horticulture in the country has been thoroughly involved with the cultivation of fruits in India, besides also laying simultaneous stress upon vegetables and gardening of rare plants. This very study of fruit and vegetable production is an arena, which provides sizeable subject of enormous scope, if pursued passionately in order to take India to great heights to cultivating fruits in a strategic manner. Cultivation of fruits in India and in the international scenario, involves the consolidation of wide spectrum of disciplines. As the new technologies and developments have gradually become readily available in the country, the cropping and cultivating systems and production practices have also remained witness to significant metamorphoses. Indeed, if dug deep into the fascinating history of fruit cultivation in the country, it can verily be seen that the Silk Route had played a key role in the ushering of various kinds of these delicious sweet and fleshy things called 'fruits' since pre-Christian era, especially from countries like China and Egypt.

Fruit cultivation in India is one such major commercial and business sector for exporting merchandise and shipping, from which much of the international revenue is incurred. Since Independence, the country has been trying to come to terms with the dazzling prospects of exporting commercial business and the land being essentially agrarian and rural, possesses ample scope for lands for cultivation and farming. However, the erstwhile tools and implements of the Central Indian administration, is being upgraded every fiscal year, a domain which truly looks towards guaranteed future. Numerous kinds of precepts, rules and practices that were common a few years ago in India, may just not be no longer be current.

India has been perhaps justly renamed as the 'fruit and vegetable basket of the world', a factor that weighs fascinatingly upon the cultivation of fruits in the country. India serving as home to a supreme variety of fruits and vegetables, holds an exclusive position in production figures amongst other countries. More than 90 percent of India's exports in fresh products travels to west Asia and East European markets. India's exports of fresh fruits and vegetables has amplified by 2437.12 crore rupees in the 2007-08 financial year. This amplification list merrily incorporates the products like walnut, fresh mangos, fresh grapes and umpteen other fresh fruits and vegetables.

Since prehistoric times, India has served as an agriculturally dependant country. Hundreds of fruits and vegetable types are grown in all parts of India. For such a thriving business to network itself, the Indian industrial strategy has been known to have expanded from grassroots level, onto big markets, dealing globally. The cultivated fresh fruit and vegetable reach small scale fruits and vegetables suppliers, they are then sent to local markets as well as fruits and vegetables exporters, who in turn make baskets of shipping of fruits, dealing with international buyers.

The last few decades have witnessed the number of Indian fruit vegetables suppliers and fruits vegetables exporters rising to an all time high. The total production of fruits and vegetables in the world is approximately 370 metric ton. India proudly ranks first in the world, with an annual output of 32 metric ton, taking the fruit cultivation of Indian graph shooting sky-high. While there are almost 180 families of fruits that are cultivated all over the world, citrus fruits represent approximately 20 percent of world's total fruit production. Major Indian cultivated fruits consist of mango, banana, citrus fruits, apple, guava, papaya, pineapple and grapes. The fruits are further also processed into various products such as fruit juices and concentrates, canned fruit, dehydrated fruit, jams and jellies etc.

India with its current production of approximately 32 million metric ton of fruits, accounts for about 8 percent of the world's fruit production. The diverse agro-climatic zones present in the country, makes it possible to practically grow almost all varieties of fresh fruits and vegetables in India. The fruit cultivation and production in India has recorded a growth rate of 3.9 percent, whereas the fruit processing sector has grown at approximately 20 percent per annum. However, the growth rates have been expansively higher for frozen fruits and vegetables (121%) and dehydrated fruits and vegetables (24 percent). There exist more than 4000 fruit processing units in India, with an combined capacity of more than 12 lakh metric ton (less than 4 percent of total fruits produced). It is reckoned that approximately 20 percent of the production of processed fruits is intended for exports, the rest supplies to the defence, institutional sectors and household consumption; fruit cultivation in India however depends a lot upon mango and mango-based products, constituting 50 percent of exports.

The fruits and vegetable cultivation and processing industry in India is highly decentralised in content. A large number of units are in the cottage/home scale and small scale sector, possessing small capacities upto 250 tonnes per annum. But big Indian and multinational companies in the sector possess enormous capacities in the range of 30 tonnes per hour more or less. Since liberalisation and pulling out of excise duty on fruit and vegetable products, there has been substantial rise in the growth rate of the industry.

The Indian fruit cultivation, especially mangoes and bananas, are now in huge demand outside the country. India is a producer of tropical fruits like coconuts, jackfruits, cashew nuts, pineapples, bananas and oranges. Of temperate fruits, apples, plums, peaches, almonds, apricots and grapes are grown in abundance. While Jammu and Kashmir and Himachal Pradesh lead in the fruit production of the temperate region, others are grown in various parts of peninsular India and Northern Plains. India brings in pots of foreign exchange by exporting cashew nuts too. Part of the raw cashew nuts is imported and treated here, before they are later re-exported. The north eastern region of India holds mammoth horticulture potential. States like Arunachal Pradesh, Nagaland, Meghalaya and Manipur have favourable soil and climatic conditions and also have great scope for temperate fruit cultivation in these parts of India to augment business management. Presently stone fruits like peach, plum and apricot and pome fruits like apple and pear are grown in small scale but the future is bright and there is great scope for the expansion of temperate fruit culture.

Fruit cultivation in India has received additional impetus in the format of various state governments realising their functionary role in boosting c nationwide commerce. As such, the Haryana government has been lending grants for setting up orchards under the National Horticulture Mission. Under this plan, a special grant of 9,750 rupees per hectare is provided to the farmers for cultivation of guava, 15,750 rupees per hectare for mango, 11,250 rupees for ber and 18,525 rupees for cheeku. This amount is to be disbursed in three years time, according to sources. Ample investment opportunities does exist in expanding the export market for fruit cultivation in India. A modifying acceptance of new products with market development efforts have also been witnessed since long, given the fact that there exists a good international demand for certain fruits and vegetable products. India produces 41 percent of world's mangoes, 23 percent bananas, 24 percent cashew nuts and 36 percent green peas. The total export value of the main exporting fruit crop from India is mango. Exports of mangoes, grapes, mushrooms have started going to the United Kingdom, Middle East, Singapore and Hong Kong.

Mango, termed as the 'king of fruits in India', accounts for 40 percent of the national fruit production almost every year. It occupies 42 percent of the country's 2205.6 thousand hectares land under fruit cultivation in the Indian subcontinent. India had exported 54350.80 metric ton of fresh mangoes with the value of 127.42 crore rupees in 2007-08.

The major varieties of mangoes exported include-Dashehri, Alphonso, Kesar, Banganpalli, Kesar, Dusheri, Langra, Chausa, Mallika and. Swarnrekha. The major markets for Indian mangoes comprise U.A.E, Bangladesh, U.K, Saudi Arabia and Nepal. India is verily approximated to account for around 60 percent (9.5 million tonnes) of the world's mango production of 15.7 million tones. The major cultivation and production areas of this fleshy fruit in Indian subcontinent are in the states of Andhra Pradesh, Uttar Pradesh, Karnataka, Bihar, Gujarat and Maharashtra. India is one of the leading exporters of fresh table grapes to the global market; India's export of grapes has increased from 301.92 crore rupees in 2006-07 to 317.83 crore rupees in 2007-08. The leading destinations for export of Indian comprise Netherlands, U.K, U.A.E, Bangladesh and Belgium.

As has already been stated that the country is the largest producer of fruits in the world, acknowledging it as the fruit basket of world, India's fruit cultivation has been orchestrated after several application of masterminds to have thus reached such a position. The major fruits cultivated in India are mangos, grapes, apple, apricots, orange, banana, avocados, guava, lichi, papaya, sapota and water melons. This is due to its potential in different agro climatic zones. India's export of fresh fruits to the major countries consists of U.K, Netherlands, U.A.E, Russia, Bahrain, Qatar, Kuwait, Saudi Arabia, Bangladesh and Nepal.

Fruit cultivation in India further has always remained a recipient of huge respect and admiration, owing to its marvellous produces of deciduous fruits, including pome fruits (apple and pear) and stone fruits (peach, plum, apricot and cherry) in considerable quantities. Deciduous fruits and its cultivation in India mainly takes place in the north Indian states of Jammu and Kashmir, Himachal Pradesh and in the Uttar Pradesh hills. The northeastern hills region, comprising the states of Arunachal Pradesh, Nagaland, Meghalaya, Manipur and Sikkim, also grow some of the deciduous fruits on a restricted scale. Due to initiation and adaptation of low chilling cultivars of crops like peach, plum and pear, they are also presently being grown commercially in certain areas of the north Indian plains too. Out of all the deciduous fruits, apple is the most vital in terms of production and extent.

From amongst the umpteen and healthy and palatable deciduous fruit cultivation in India, apple demands and calls for incredible attention. Indeed, apple was introduced into the country by the British Raj in the Kullu Valley of the Himalayan state of Himachal Pradesh as far back as 1865. On the other hand, the coloured 'delicious' cultivars of apple were introduced to the Shimla hills of the same state in 1917. The apple cultivar Ambri, is measured as indigenous to Kashmir and had been grown long before Western introductions. Pears and other deciduous fruits were domesticated lucratively during the early part of the 20th century, although some of them were accounted to occur under semi-wild conditions much earlier. Apricot was found growing in the drier pockets of northwestern Himalayas and two apricot varieties, locally acknowledged as Halman and Rakchaikarpo, are reported to be indigenous to the Leh-Ladakh region Jammu and Kashmir state.

Sweet cherry was introduced from Europe before India's independence in 1947, whereas, commercial cultivars of sour cherry have been fetched primarily from USA in more recent years. The European and Japanese plum kinds are cultivated both in high and low hill areas. A plum variety 'Santa Rosa', reported to be a hybrid between Japanese and American species, predominates (70-80 percent) plantations in the Himalayan hills. Low chilling cultivars of peach and nectarine such as Flordasum, Flordared, and Sunred nectarine are booming introductions to the north-Indian plains. Some local selections of peach, plum and sand pear (Patharnakh) have also taken fruit cultivation in India on a grand commercial scale in subtropical marginal chilling areas of north India.

Fruit cultivation in India, quite understandably grows most prolific in the areas of the Himalayas, with abundance speaking out from every angle. The Uttar Pradesh hills, particularly the Kumaon hills division, possesses exceptional advantage of early harvest of apple, principally due to cultivation of early maturing varieties like Early Shanburry, Fanny and Benoni. The early maturing varieties are harvested 2-3 weeks before the onset of fresh apple from Himachal Pradesh and Jammu and Kashmir and hence, fetch very moneymaking prices. Deciduous fruits, covering pome and stone fruits contribute significantly to the horticulture economy of India. Truly, cultivation of fruits in India, does depend upon the success of plantation, harvest, soils and a perfect ambience of brilliant minds.

Deciduous Fruit Production in India

India produces all deciduous fruits including pome fruits (apple and pear) and stone fruits (peach, plum, apricot and cherry) in considerable quantity. These are mainly grown in the North-Western Indian States of Jammu and Kashmir, Himachal Pradesh and in Uttar Pradesh hills.

The North-Eastern Hills region, comprising of the States of Arunachal Pradesh, Nagaland, Meghalaya, Manipur and Sikkim also grows some of the deciduous fruits on a limited scale. Due to introduction and adaptation of low chilling cultivars of crops like peach, plum and pear, they are also now being grown commercially in certain areas of the north Indian plains. Out of all the deciduous fruits, apple is the most important in terms of production and extent.

Apple was introduced into the country by the British in the Kullu Valley of the Himalayan State of H.P. as far back as 1865, while the coloured 'Delicious' cultivars of apple were introduced to Shimla hills of the same State in 1917. The apple cultivar 'Ambri', is considered to be indigenous to Kashmir and had been grown long before Western introductions. Pears and other deciduous fruits were domesticated successfully in the early part of the 20th century, although some of them were reported to occur under semi-wild conditions much earlier. Apricot was found growing in the drier pockets of northwestern Himalayas and two apricot varieties, locally known as 'Halman' and 'Rakchaikarpo' are reported to be indigenous to Leh-Laddakh areas of J&K State.

Sweet cherry was introduced from Europe before India's independence in 1947, while commercial cultivars of sour cherry have been brought mainly from USA in more recent years. The European and Japanese plum varieties are grown both in high and low hill areas. A plum variety 'Santa Rosa' reported to be a hybrid between Japanese and American species predominates (70-80%) plantations in the hills. Low chilling cultivars of peach and nectarine such as Flordasum, Flordared, and Sunred nectarine are successful introductions to the north-Indian plains. Some local selections of peach (Shan-e-Punjab, Sharbati), plum (Jamuni, Alubhokhara) and sand pear (Patharnakh) are also cultivated on a commercial scale in subtropical-marginal chilling areas of north India.

Present Situation of Deciduous Fruit crop Cultivation

Crops Grown

India grows crops like apple (*Malus pumila Mill.*), pear (*Pyrus communis Berm.f.*), peach/nectarine (*Prunus persica (L). Bats. Ch.*), plum (*Prunus domestica L.*), apricot (*Prunus armeniaca L.*), sweet cherry (*Prunus avium L.*) and sour cherry (*Prunus cerasus L.*) on a commercial scale.

Cultivars

The promising cultivars of different temperate fruits in 3 major deciduous fruit growing States of India. Some relevant information on cultivars and cultivar selection is indicated below:

Apple: Over 700 accessions of apple, introduced from USA, Russia, U.K., Canada, Germany, Israel, Netherlands, Australia, Switzerland, Italy and Denmark have been tried and tested during the last 50 years. The delicious group of cultivars predominates the apple market. The areas covered under Delicious cultivars are: 83% of the area under apple in H.P., 45% in J&K and 30% in U.P. hills. In more recent times improved spur types and standard colour mutants with 20-50% higher yield potential are favoured. The important selections are:

- Spur types-Red spur, Starkrimson, Golden spur, Red Chief and Oregon spur.
- Colour mutants-Vance Delicious, Top Red, Skyline Supreme.
- Low chilling cultivars-Michal, Schlomit.
- Early cultivars-Benoni, Irish Peach, Early Shanburry, Fanny
- Juice making cultivars-Lord Lambourne, Granny Smith, Allington Pippin.
- Scab resistant cultivars-Co-Op-12, Florina, Firdous, Shirean.
- New Hybrids-Lal Ambri (Red Delicious X Ambri), Sunehari (Ambri X Golden Delicious), Amred (Red Delicious X Ambri), Chaubatia Anupam & Chaubatia Princess (Early Shanberry X Red Delicious) developed in India.

In H.P. monoculture of a few cultivars such as Royal Delicious, Red Delicious and Richared have started showing negative impact on the apple industry. Serious problems of apple scab disease and outbreak of premature leaf fall and infestation of red spider mite are causing great concern. U.P. Hills, particularly the Kumaon hills division, have the unique advantage of early harvest of apple, mainly due to cultivation of early maturing varieties like Early Shanburry, Fanny and Benoni. The early maturing varieties are harvested 2-3 weeks before the arrival of fresh apple from H.P. and J&K, and hence fetch very remunerative prices.

Pear: In pear for higher altitude conditions high chilling requirement varieties (like Bartlett) are mainly grown. In more recent years, red-colour strains of pear like Max Red Bartlett, Red Bartlett and Starking are replacing yellow coloured cultivars. In warmer sub-mountainous areas of H.P. and subtropical Punjab oriental pear cultivars like Baghugosha, Kieffer, China and sand pear Patharnakh are cultivated commercially both for table and processing purpose.

Apricot: Generally there are two types of apricot, namely, sweet kernel type and bitter kernel type. About 81 exotic accessions and 19 indigenous cultivars were collected and evaluated. The local types Halman and Rakhaikarpo have been popular, whereas exotic introductions namely, Nari, Kaisha, Shakarpa and New Castle are promising. These cultivars are recommended for dry cold areas. The USA variety Nugget is self-fruitful and bears sweet and attractive coloured fruits.

Peach: For colder conditions the peach cultivars July Elberta, Elberta, Peshwari, Quetta, Burbank and Stark Earliglo are well adopted. Low-chilling cultivars viz. Flordasum, Flordared, Shan-e-Punjab, Sharbati and Sunred (nectarine) have become popular in subtropical belts of U.P. and Punjab States.

Plum: A large number of cultivars have been introduced into the country. European plums performed better in the hills, while Japanese plums are more adopted in sub-mountainous lower elevations. Leading cultivar in the hills is Santa Rosa. In the north-Indian plains small fruited cultivars like Titron, Kala Amritsari, Kelsey, and Alubukhara showed better performance. A good number of low-chilling Florida hybrids Sungold, Redgold etc, are under evaluation.

Cherry: Many cultivars of sweet cherry have been introduced from Europe, USSR and British Columbia. Promising exotic cultivars like Bigarrean Napoleon, Black Heart, Guigne Noir for J&K and cultivars like Black Tartarian Bing, Napoleon (white) Sam, Sue (White), Shella for H.P. have been identified. For warmer climate, cultivars like Summit, Sunburst, Lapins, Compacat and Stella have been found to be promising.

Rootstocks

- Apple has been propagated mainly on seedling stocks of apple (*Malus pumila Mill*). Clonal rootstocks of M and MM series were introduced from East Malling Research Station of U.K. and M-9, M-26, M-4, M-7, MM-106 and MM-111 were identified as promising. Merton 779 was recommended as a commercial rootstock for apple for Kumaon hills of U.P. However, clonal rootstocks have not found favour among the apple growers. *In-vitro* protocols of M-9, MM-106 and MM-111 have been developed. Clonal rootstocks have been preferred only for high density orchards. Four rootstocks, namely, EMLA-111 for drought prone areas, EMLA-7 for slopy land and EMLA-106 for slopy and less clayey soil and EMLA-9 for high density planting on irrigated deep soils have been recommended. In pears, seedling stocks of *Pyrus pashia* (Kainth) is commercially used. Clonal rootstock Quince A with semi-vigorous growth has some promise. Newer clonal rootstocks like Old Home X Farmingdale series and BA 29C are under evaluation.
- For apricot, seedlings of wild apricot have been found to be good giving a smooth union and vigorous scion growth. Peach can also be tried as a rootstock, particularly for dry and light soils. Myrobolan plum is recommended for high moisture soil conditions.
- Peach is normally grafted on wild peach seedlings, plum and even on apricot. Wild peach produced healthy and high-yielding plants. For controlling tree size, apricot rootstocks which impart dwarfing may be used.

- For plum, wild peach and wild apricot are commonly used as rootstocks. Among clonal rootstocks, Myrobolan B. was found to be most promising. Wild apricot seedlings and wild peach were also found to be suitable rootstocks for commercial cultivars like Santa Rosa, Mariposa and Beauty.
- Commercial plantations of sweet cherry were raised mainly on Mazzard and Mahaleb rootstocks. Seedlings are raised from late maturing cherry cultivars and stones of Waterloo or Misri cultivars are often used as rootstocks. Rooting of cherry suckers has been made possible with use of rooting hormones like IBA (6000 ppm) in combination with NAA (2000 ppm).

Area and Production

Deciduous fruits are mainly cultivated in North West Hills Region of India, comprising of States of Jammu & Kashmir (J&K), Himachal Pradesh (H.P.) and Uttar Pradesh (U.P.) and in the North Eastern Hills Region in the States of Arunachal Pradesh, Nagaland, Meghalaya and Manipur. Available data on area and production of apple and pear are shown in Table.

Table. Area, Production and Yield of Apple and Pear in India

State	Apple (1995-96)			Pear (1992-93)		
	Area (ha)	Production (MT)	Yield (MT/ha)	Area (ha)	Production (MT)	Yield (MT/ha)
Arunachal Pradesh	5523	9730	1.76	938	2035	2.17
Himachal Pradesh	78296	276681	9.40	4653	8767	1.24
Jammu and Kashmir	78007	714834	11.0	14012	17941	1.28
Nagaland	64	109	1.70	460	1480	3.22
Uttar Pradesh	55200	210000	3.95	10550	27000	2.56
Total	217099	1211379	5.56	38536	175808	2.09

Apple

Although there has been 5-6 fold increase in apple production during the last 50 years, the productivity level is still very low (5.56 t/ha). Apple cultivation received greater attention by the growers. In H.P, area under apple increased from 3026 ha in 1960-61 to 78296 ha in 1995-96 with a corresponding increase in yield. J&K covers about 78007 ha under apple with a production of 714834 tons. In the U.P. hills (8 districts) apple occupies about 30 per cent of the area under fruits and contributes 46.9 percent of fruit production.

The area covered under apple in U.P. hills is 55200 ha with production of 210000 tons of fruits. In the North-Eastern Hills Region, good quality apple is produced only in the rain-shadow belts of Arunachal Pradesh (5523 ha), and in Nagaland a very small area (64 ha) has been brought under apple cultivation.

About 99 percent of India's apple area falls under the North Western Hills region, covering 6 districts of J&K (Srinagar, Budgam, Pulwama, Anantanag, Baramullah, Kupwara), 6 districts of H.P. (Shimla, Kullu, Sirmour, Mandi, Chamba, Kinnaur) and 8 districts of U.P. (Almora, Nainital, Pithauragarh, Tehri, Pauri, Chamoli, Uttarkashi, Dehradun). In the Northeastern Hills region, good quality apple is grown in a small area in Tawang belt of Kameng district in Arunachal Pradesh. The Tawang area is basically a rainshadow belt and therefore, permits a longer period of sunshine and freedom from heavy rains, making it ideal for apple. Apple is also grown in Sikkim and Nagaland but the production is not a major success. Presently, a small quantity of apple produced in India is exported, mainly to Bangladesh and Sri Lanka.

Pear

Pear is grown under a wide variety of climatic regimes, ranging from cold dry temperate hilly conditions to warm humid subtropical conditions on the plains of northern India. Its cultivation is most extensive in J&K State (14012 ha), followed by U.P. (10550 ha). In the plains of Punjab, pear is grown in a specific area (7899 ha), whereas in H.P, it is widespread from the foothills to the highlands (600-2700 m) in areas experiencing 500-1500 chilling hours. Area, production and yield of apple and pear.

Others

There is no reliable data on area and production of other deciduous fruit crops. Stone fruits as a group occupy an area of 0.11 million ha, with 0.14 million tons production. In H.P about 28 thousand ha are covered under peach, plum, apricot, almond and cherry. In the Northwestern Himalayan region, peach holds greater promise because of its utilization for canning purposes. Peach is grown mainly in low and mid hilly areas (1000-2000 m above msl), except the low chilling cultivars belonging to the Florida group, which can be grown very well under subtropical conditions.

European plums require more chilling than Japanese plum. Japanese plums are generally grown in low and mid-hilly areas (1000-6000 m) conditions. The climatic requirement of apricot is almost the same as that of plum, while cherries require a colder climate (1000 chilling hours below 7° C). Cherry is commercially cultivated in J&K where the hailstorm problem is not encountered.

In the North-Eastern Hills (NEH) Region, these crops can be successfully grown in Arunachal Pradesh (Kameng, Siang, Tirap and Lohit districts), Meghalaya (Central plateau of Khasi & Jaintia hills), Manipur (Maram, Tadubi, Mou, Ukhrul) and Nagaland (Mokokchung, Wokhla, Thensang, Kohima, Phek districts). The productivity of all the stone fruits is low and estimated yield of peach is 0.50, 0.73, 2.11 and 2.23 t/ha in H.P, J&K, U.P, and NEH Regions respectively. The productivity of cherry in J&K is approximately 1.73 t/ha, while average yield of apricot is 0.42, 0.20 and 0.28 t/ha in H.P, J&K and U.P. hills respectively.

Production of Planting Material

Nurseries

There is a large number of government and private nurseries engaged in multiplication of planting material of deciduous fruit crops. In addition, the State Agricultural Universities and the Research Institutions multiply planting material of improved cultivars for sale and distribution to the farming communities. The Government of India had supported establishment of a large number of fruit nurseries, both in public and private sectors as planned activities. During the 8th Five Year Plan (1992-97) it is estimated that over 55 million nursery plants of different perennial crops, including temperate fruits, have been produced and distributed under the scheme. There was a target for the establishment of 85 big nurseries, 587 small nurseries, and 37 tissue culture units (20 by the private sector and 17 by the Government) of different fruit crops. Exact number of these nurseries engaged in production of pome and stone fruits is not known. However, the existing nurseries (more than 600), covering both public and private Institutions, are sufficient to meet the requirement of planting material of deciduous fruit crops.

Propagation and Rootstock Information

In all the pome and stone fruits vegetative propagation techniques of budding or grafting are followed for multiplication of planting material on standard rootstocks, raised both from seeds or through clonal methods. Cropwise details are as follows:

Apple

For raising rootstock seedlings, seeds of crab apple or commercial cultivars are stratified during December for 2-3 months at 2-5°C. One year old seedlings are used for budding/grafting. Clonal rootstocks are raised through mound or stool layering. The mother plants are allowed to grow for one season and cut back to 3 cm from the ground level just before the growth begins. When new growth is about 10 cm, the shoots are covered with soil leaving the growing parts exposed. Rooted layers are cut off close to the ground level and planted in nursery beds for grafting/budding.

In H.P. hills, different types of budding/grafting are recommended:

* Chip budding in mid-June and mid-September
* Whip and Tongue and cleft grafting in February-March
* T-budding in May-June.

Pear

Seedlings of Kainth (*Pyrus pashia Linn.*) or Shiara (*P. serotiana Rehd*) are stratified in moist sand for 35-45 days at 2-5°C. Tongue grafting in February-March or T-budding during June-July is recommended.

Others

Plum and apricot are generally grafted on wild apricot seedling rootstocks. Peach is also used as a rootstock. Peach is budded/grafted on wild peach and peach-almond hybrid rootstocks. Hard wood cuttings (treated with 500 ppm IBA) can also be employed for raising rootstocks.

For apricot, peach, plum and cherry shield budding during July-August has been recommended for the hills of J&K.

Micro-propagation through tissue culture technique has not yet been commercialized for temperate fruits, although tissue culture protocols have been developed at research Institutes/Universities. Colt, a rootstock for cherry can be multiplied in large numbers by following the tissue culture technique. Shoot buds are cultured in MS medium containing 1-2 mg/litter benylaminopurine. Individual shoots are rooted on MS medium containing 1 mg/litter Indole Butyric Acid (IBA). The rooted plants are transferred to sterilized sand + soil (1:1) mixture and covered for 7-10 days for hardening.

Establishment of Orchards

Land Preparation and Planting

Fertile sandy-loam to clay loam soils with pH range from 5.5 to 7.5 and free from water logging conditions are suitable for establishing deciduous fruit orchards.

In flat lands/valleys, square, rectangular or triangular planting may be adopted. In hill slopes, planting on contours or terraces is recommended. In shallow slopy lands, small terraces (half-moon terrace), may be made to establish the plants and large scale disturbance of surface soil need to be avoided.

Planting of deciduous fruit plants is done during winter months from the end of December to mid March. Planting is done in the centre of pits (1x1x1m cube or circle), prepared a month before planting. While refilling the pit, 50g of aldrin powder (15%) is mixed with the soil. After planting, the young plants are supported with stakes and basins are kept free of weeds. Mulching with dry grass or polyethylene is advisable. Irrigation is provided after planting.

Spacing

Spacing varies from species to species and depends also on the type of rootstocks used. For apple raised on seedlings of crab or other commercial varieties of apple, a planting distance of 6x6m or 7x7m accommodating 277-205 plants per ha is recommended for J&K hills. On clonal rootstocks like M4, M7, M26, MM106 a spacing of 4.5m x 4.5m (555 plants/ha) is suggested.

For spur type varieties and standard coloured mutants of apple, high density planting on dwarfing rootstocks like M9, M4, M7 and MM107 has been found to be feasible. Fruit yield of 30-35 tons/ha has been achieved in 12 year old orchards of colour mutants of apple on MM106 under a planting density of 2222 plants/ha (3m x 1.5m) in cooler hills of H.P. Spacings recommended for apple are shown in Table.

Table : Optimum Plant Spacing for Spur Types and Standard Colour Mutants of Apple

Type	*Rootstock*	*Spacing (m×m)*	*No. of trees/ha*
Spur Type	Seedling (Crab)	5×5	400
Spur Type	MM111, MM109	4×4	625
Standard Type	MM106, MM109	5×5	400
Spur Type	MM106, M7	3×3	1111
Standard Type	M9	2×2	2500

Source: Awasthi, R.P. and Chauhan, P.S., 1997. Apple and Pear. In: 50 years of Crop Science Research, ICAR Publication, New Delhi.

In pear, for trees on seedling rootstock under normal conditions, spacing of 5x5m is recommended, but on clonal rootstocks (Quince A) spacing can be reduced to 3x3m. In peach, a spacing of 4.5 to 5m and for plum, apricot and cherry a spacing of 6m are followed.

Care and Management of Orchards

Training and Pruning

Modified central leader system of training has been recommended for both apple and pear on seedling rootstocks. The proportional heading back and thinning out system of pruning is adopted after the juvenile phase of plant growth. Spur pruning encourages vegetative growth and helps in new spur development in old plantations. For high density planting on semi-dwarfing and dwarfing rootstocks spindle bush, dwarf pyramids and cordon system of training are suggested. Spindle bush on M7 and modified central leader on MM106 rootstocks are successful.

Peach is generally trained to open centre system, while in sweet cherry modified central leader system is followed. Pruning is done in such a way so that 20-50 cm new growth in young trees and 25-30 cm in older trees are secured every year. In sweet cherry about 10% of fruit bearing area should be removed annually, whereas in apricot slightly heavier pruning should be practiced to keep the spur system renewed.

Manures and Fertilizers

Nutrient requirement varies from place to place. In Himachal Pradesh widespread deficiencies of N, P, K, Ca, Mn, Zn and B have been recorded. A fertilizer dose of 700:350:700 g of N, P and K for full grown bearing apple tree has been recommended.

It has been observed that VAM (Vesicular arbuscular mycorrhizea) fungi increases P uptake by apple roots. The following corrective measures for nutrient deficiencies have been suggested for apple.

Table : Corrective Measures for Nutrient Deficiencies in Apple

Elements	*Chemical & Dose*	*Time of Spray*
N	Urea, 5.0%	Pre-petal fall
Ca	Ca Cl_2, 0.5%	30-45 days before harvest
Zn	$ZnSO_4$, 0.5%	After petal fall
Mn	Mn SO_4, 0.4%	After petal fall
B	H_3 BO_3, 0.1%	Before bloom and after bloom

In peach cv. Sharbati, 17.5 g N/tree/year and in apricot 70-100 g N/tree/year have been found to be optimum. For peach, N has been the main limiting element in H.P, while for plum, maximum fruit yield was recorded when leaf N, P and K contents were 2.89, 0.28 and 0.89 respectively.

Weeding and Mulching

In apple orchards, grass mulching (10cm thick) with one application of post-emergence herbicide Glyphosate (0.8 kg/ha) was found to be effective. The shrubby weeds were, however, best controlled by 500 ppm Gramaxone + 100 ppm 2,4,5-T application. Moisture conservation was maximum under grass mulching and, therefore, recommended for apple in H.P. Mulching with oak leaf was effective in U.P.

Supplementary Irrigation

Trickle irrigation at 75 per cent of field capacity results in better tree growth and higher fruit yield in apple when raised on semi-dwarfing rootstocks. For apple, most critical period of water requirement is April to August and peak water requirement is just after fruit set. 114 cm of water during whole year through 19 irrigations have been recommended. Drip irrigation saves water considerably. Field trials indicated a total irrigation requirement of 3840 liters water per tree under conventional system of irrigation; under drip system 1695 liters of water was enough. The application efficiency under drip system is about 2.27. In Santa Rosa plum, irrigation at 50% field capacity gave better growth and economic returns.

Pest and Disease Control

Pests

In apple and pear about a dozen pests are causing serious damage to the crops. Most important ones are: San Jose Scale, Woolly apple aphid, Root borer, Blossom thrips, Codling moth and European red mite. San Jose Scale can be effectively controlled with eco-friendly miscible spray oils at 2% concentration when applied during

February-March. Woolly aphids can be controlled through soil application of Phorate or Carbofuran granules during May and October/November. For codling moth, pheromone trapping was found to be effective, and certain bio-control agents have also been identified in apple orchards for controlling certain insect pests.

European red mite is becoming a very serious pest for apple in H.P. and J&K. The pest attack causes premature leaf fall. Late dormant sprays of miscible oil provide effective control of eggs. The mite can also be controlled by sprays of Dicotol (0.05%) followed by Malathion (0.05%).

In peach, leaf curl aphid can be controlled by pre-bloom sprays (at pink bud stage) of Dimethoate (0.03%) or Monocrotophos (0.04%). Fruit fly in peach, apricot and plum can be controlled through foliar spray of baits consisting of Malathion (0.1%) + 1% sugar. Stem and shoot borers causing damage to peach, plum, apricot, almond and cherry can be controlled by inserting 0.5g of PDCB (paradichlorobenzene) into the holes and plugging them with mud.

Diseases

Apple scab caused by *Venturia inaequalis* is a serious disease causing maximum economic loss. A sound forecasting and early warning system has been developed for prediction of scab attack. Also, a judicious fungicide spray schedule has been devised. Under high disease pressure, systematic fungicides performed better, while under low disease pressure Ergosterol biosynthesis inhibiting (EBI) fungicides were as good as protectants. Ascosporic inoculum produced by over-wintered apple leaves could be substantially reduced by giving post-harvest applications of Bavistin (0.1%) and EBI chemicals Penconazole (0.5%) and Flusilazole (0.01%) as preharvest fungicidal sprays control scab during storage.

Powdery mildew disease can be kept under check by pruning and spraying wettable sulphur (0.2-0.3%) or Karathane (0.05%) during dormancy, bud swell and petal fall stages.

Collar rot and white root rot diseases in apple occur mainly in poorly drained soils. Proper drainage of orchards and soil drenching with 0.1% Carbendazim for white root rot and with 0.3% Mancozeb or 0.3% Ridomil MZ for collar rot are effective.

Virus and virus-like diseases such as mosaic, chlorotic leaf spot, rubbery wood and others have been reported. Virus cleaning through tissue culture and supply of virus free bud wood material are being pursued to contain further spread of the viral diseases in apple.

Intercropping

As apple is grown in hill slopes, seasonal intercropping accompanied by working the soil is not encouraged in order to prevent soil erosion. Where slope of the land is more than 10 per cent, various grasses and legumes are grown as permanent covers.

The tree basins are kept clean and sod culture in the orchards is practiced in the hills of H.P. The grasses are regularly cut when 7-10 cm high and the residue is left in the orchards for decomposition.

The cultivation of legumes and grasses as intercrop is becoming popular in orchards. The important legumes are red clover (*Trifolium pratense*), white clover (*T. repens*) and lucern (*Medicago sativa*). Among the grasses, orchard grass (*Dactylis glomerata*), fescues (*Festuca arundiceae*) and Timothy (*Phleum pratense*) are common. Orchard grass is grown where moisture is high, while fescues is suitable for dry areas or southern aspects. In apple, sometimes filler trees of peaches or dwarf apple plants are grown.

At the young age of orchard plantations, vegetables like cole crops, potato and tomato are grown to supplement the income. However, this practice is abandoned when the orchard starts bearing after 5-7 years to avoid competition. In Shimla hills and Kullu Valley areas of H.P. cultivation of gladioli for flower and bulb production as intercrop is carried out in the apple orchards since it can tolerate shading.

Harvesting and Yield

Maturity Standards: In apple, the number of days taken from full bloom to harvest are 132-134 days for Starking Delicious, 138-140 days for Red Delicious and 147-148 days for Golden Delicious varieties in the high hills (above 2000 m) of H.P. In J&K State, the variety Maharaji requires 160 ± 3 days after full bloom for proper harvest maturity. Apple cultivar Granny Smith takes 180 ± 5 days for maturity. The maturity standards for different cultivars of apple are given in Table.

Table : Maturity Indices for Commercial Cultivars of Apple

Cultivar	*Days from full bloom*	*Firmness (kg)*	*TSS (%)*
Royal Delicious	125 ± 5	8.2 ± 0.40	13.0 - 15.0
Red Gold	122 ± 3	8.3 ± 0.20	12.0 - 13.5
Red Delicious	134 ± 5	8.4 ± 0.40	10.0 - 14.0
McIntosh	135 ± 4	6.8 ± 0.25	11.5 - 13.5
Golden Delicious	148 ± 6	8.4 ± 0.40	12.0 - 14.5
Granny Smith	180 ± 5	8.7 ± 0.30	11.5 - 13.0

Fruit Drop Control

Pre-harvest fruit drop in apple could be effectively checked with the application of 10 ppm NAA 20-25 days before harvest. Pre-harvest fruit drop is quite serious in early ripening cultivars like Tydeman's Early, Red Gold and Pippins, where loss of 40-60% crop load is often experienced. Mid-season Delicious group cultivars also experience 15-20% loss.

Colour Development and Enhancement of Ripening

In apple, fruit colour development in warmer and lower (below 1800 m) elevations is generally poor. Application of Ethephon (2-chloroethyl phosphonic acid) about 10 days before harvest improves red coloration. Induction of early maturity by 7-10 days in mid hills situations could be achieved through an application of 500 ppm of ethephon + 10 ppm of NAA 3 weeks before harvesting. In H.P, two pre-harvest sprays of 0.5% $CaCl_2$, followed by a 0.03% surfactant like Tween 20 for 1-2 minutes have been recommended for better shelf-life of apple.

Fruit Thinning

In apple, for obtaining better grade fruits of optimum size a post-bloom spray of 100-200 ppm ethephon 2-3 weeks after full bloom has been recommended both in H.P. and J&K.

Pear

Harvesting time of pear is determined by slight change in fruit colour and easy separation of fruits. Fruits should not be allowed to ripen to soft condition on the tree and should be harvested when they are still hard. Depending on cultivars fruit may be ready for harvest from 70 days after full bloom to 135 days after full bloom. The Bartlett pear normally takes 122 ± 3 days in valley areas of H.P.

Cherry

Sweet cherry should be harvested when fruit stalk separates easily from the spur and taste develops typical to the cultivar. From full bloom it takes 45 days in cultivars like Guigne Noir and Gross Lucenta, whereas in cultivars like Bigarrean Napoleon the number of days varies from 65-75 days.

Peach

For harvesting peach at the right time, the proper colour development in fruits and pit browning are considered as reliable guides. In yellow fleshed cultivars, deep orange colour development on fruits is associated with proper maturity. At least 5% of the pit area also should be brown if the fruit is to develop good flavour at ripening. In the peach variety July Elberta, it takes about 90-95 days after full bloom, whereas on cv. Elberta it normally takes 100-105 days. In low chilling cultivars like Flordasun, fruits should be harvested at 50% colour development stage.

Plum

In free-stone cultivars of plum, pit browning up to 5% of the pit area is a good index of maturity. For proper fruit maturity, early cultivars like Sharp Early and Formosa take 95 to 105 days, mid season variety Santa Rosa requires 115-120 days and late cultivars like Grand Duke take 130-135 days after full bloom.

Apricot

Apricot fruits of a single tree ripen for a period of about 3 weeks. The fruits develop detectable flavour when allowed to ripen on the tree. The early cultivar Charmagz matures within 75 days from full bloom, while the late cultivar Quetta requires 125-130 days for proper harvest maturity.

Yields

There is a wide range of variation in fruit yield in different States and in different cultivars. The average yield of apple and pear have been indicate. Although average productivity of apple in H.P. has been shown as 9.4 t/ha, some well managed orchards of Delicious apple in Kotgarh area of the same State yield about 50 t/ha. Similarly, yield record of 35 t/ha has been achieved in 12 year old plantations of coloured mutant apple cultivars on MM 106 rootstocks under high density planting system (2222 plants/ha) in a research station in Shimla hills of H.P.

The average yields of different stone fruits have been indicated earlier. In brief, State-wise average yield of different stone fruits are given in Table.

Table : Average Yield of Different Stone Fruits by State

State	*Fruit Yield, T/ha*			
	Peach	*Plum*	*Apricot*	*Cherry*
H.P.	0.50	1.31	0.42	-
J&K	0.73	0.74	0.20	1.73
U.P.	2.11	2.50	0.28	-
NEH Region	2.50	1.18	-	-

Marketing

The biggest wholesale market for apple is the Fruit and Vegetable market at Azadpur, in Delhi. About 70 per cent of the total trade of apple is distributed through this market. The seasons of market arrivals from North Western States are as follows:

H.P. - July to October with a peak in August - September

J&K - August to November, with a peak in September - October

U.P. - June to October, with a peak in July - mid September.

There are a number of marketing channels, of which the predominant are:

a) Farmer - Pre-harvest contractor - Commission agent - Wholesaler - Retailer - Consumer.

b) Farmer - Forwarding agent - Commission Agent - Wholesaler - Retailer - Consumer.

c) Farmer - Commission agent - Wholesaler - Retailer - Consumer.

The sale through pre-harvest contractors is the most important system of marketing. Normally, the small orchardists sell their crop at flowering stage to contractors who organize plant protection practices, picking and packaging of fruits.

The medium and large orchardists prefer to market their produce through channels (b) and (c). In H.P., about 65% of total apple produced is marketed through these two channels. Some Growers' Co-operatives and Government controlled marketing Corporations (like HPMC in H.P.) also get involved in the apple market.

For example, in H.P the volume of trading handled by Growers' Co-operatives is about 3.5 to 4.0 per cent, while another 2-2.5 per cent of the total produce is marketed by HPMC.

The National Horticulture Board (NHB) of the Government of India regularly publishes and announces the wholesale price and market arrival figures of apple in different terminal markets for the benefit of the growers. Some quantities of apple are placed in cold storage facilities.

The Himachal Pradesh Horticultural Produce Marketing and Processing Corporation Ltd. (HPMC) has set up modern fruit packing houses in different parts of the State. Each packing house has a capacity to grade and pack 55,000 to 275,000 cartons of fruits. Cold storage facilities at the production centres and terminal markets at Delhi, Bombay (Mumbai), Madras (Chennai) and Calcutta have been created. Similarly, the HPMC has undertaken large scale processing of temperate fruits like apples, peaches and others.

In H.P, grading and packing of fruits have been organized reasonably well. In apple, 7 size grades have been prescribed and dimensions of packing boxes, size of wrapping papers and number of layers of fruits for each grade have been standardized. In apricot, plum and peach, three size grades have been prescribed.

Presently CFB (Corrugated Fibre Board) telescopic tray pack cartons for apple and smaller size universal CFB cartons for stone fruits are in use. These are available with HPMC at their various packing stations. Plastic crates (both collapsible and non-collapsible) are also used for cold storage and for processing units.

Picking and packaging of fruits are done in different ways at different sites. In U.P hills, trees are generally strip picked into small wicker baskets or small gunny bags which are emptied into larger baskets for carrying to packing stations. Both U.P. and J&K have developed packaging units for different temperate fruits.

India presently exports a small quantity of apple (5.95% of total fresh fruit export), mostly to Bangladesh and Sri Lanka. Apple cultivars like Red and Royal Delicious, Ambri and other new coloured cultivars are suitable for the export market. According to the Trade Year statistics of 1993, India exported only 7000 tons of apple during that year.

Processing

The HPMC in Himachal Pradesh owns two fruit processing plants with a combined capacity of 30,000 tons annually. The plants are equipped with modern machinery for six-fold juice concentration and aroma recovery. The plants have also facilities for aseptic bulk packaging and pasteurization. The HPMC manufactures mainly primary products like fruit pulps, single strength juices and juice concentrates. It covers fruits like apple, pear, peach apricot and plum. The pomace of the fruits after juice extraction is used for cattle feed, extraction of pectin etc.

In J&K, low grade fruits are processed for various products like apple juice, jam, jelly, fruit bars, rings, vinegar etc. from pome and stone fruits. In U.P. hills also, various types of processed products are prepared from temperate fruits. The units are mainly small and medium in size and not to the scale of HPMC. On the research side, important cultivars of apple were screened for their suitability for preparation of juice, jam, dehydrated rings, chops, cubes and preserves. Golden Delicious cultivar was found to be the multi-use cultivar to meet the FPO quality requirements. A new process for canning of apples called osmo-canning has been developed. Similarly, process for preparation of cider, wine and vermouth have been developed. Utilization of apple pomace for production of pectin and natural colours have been attempted. Although CA storage concept has not yet become popularized, fruits placed in CA store at Shalimar in J&K have shown that this system of storage has great promise for temperate fruits.

Potential for Deciduous Fruits

Area Expansion

Both North-West and North-Eastern regions of India offer large areas ideally suitable for cultivation of pome and stone fruits. In the North-West India, stone fruits like peach, plum and apricot come up well at elevations between 900 to 1500m with an annual rainfall of 90-100cm. Apple, cherry and pear are commercially successful at elevations between 1500-2700 m above msl. The cold arid regions between 1550-3650m with annual rainfall of 25-40 cm are again highly suitable for dry nuts and drying type of apricots. There are vast tracts of land still available for further expansion of these crops.

In the North-Eastern Hills, excepting the rain shadow belts of Arunachal Pradesh and high altitude Lachung area of Sikkim, apple may not be successful commercially. Other crops like pear, peach, plum and apricot offer good scope for further expansion.

Widening the Cultivar Base

The low productivity and poor quality of apple are often linked with monoculture of a few old cultivars and their degeneration over the years. The U.P hills, particularly

Kumaon division, has the unique advantage of harvesting apple fruits for early market. Similarly, the rainshadow belts in North-Eastern Hills can offer good quality apple for the eastern Indian markets, thus reducing the cost of long distance transportation from North-Western Hills. Part of the markets of Bangladesh can be captured by the fruits of the North-East. There is good scope for introduction of new promising cultivars, replacing the Delicious group. Similarly, use of clonal rootstocks of Malling and Malling-Merton series and even their indexed material 'EMLA' selections will greatly change the productivity and quality of fruits. High density planting with spur type cultivars offer good scope. The identification of low chilling peach, plum and pear cultivars offer good possibilities for their cultivation in the low hills and in subtropical plains. Some of the new hybrids, including scab resistant apple cultivars need verification trials on a commercial scale.

Management Practices

Scientific water management and practicing proper training and pruning of trees including introduction of renewal pruning techniques, will make significant impact on increased production even in the existing orchards. Drip irrigation, *in-situ* water harvesting and correcting macro and micro-nutrient deficiencies will go a long way in bringing notable improvement in productivity as well as fruit quality. By adopting IPM strategy and organic farming practices, selected export markets can be targeted well.

Processed Products

There is immense scope for increasing various processed products of pome and stone fruits, for which technologies are available. The existing capacity of the most organized processing unit of HPMC is only marginal as compared to the volume of fruit available for processing. The HPMC utilized only 1 to 1.5 percent of total cull fruit available. The present combined capacity of two units of HPMC is 30,000 tons annually, which can be easily raised to 50 to 75 thousand tons.

The expansion of grading and packing stations, their further modernization with mechanical grading equipment, use of CFB boxes and more number of pre-cooling and cold storage units will improve the marketing system and enhance marketability of the produce. Apple has been identified as one of 6 most promising fruits for fresh fruit exports.

Constraints in Deciduous Fruit Production

Large number of old orchards (more than 30 years old) are showing decline in terms of growth and fruit yield. Such old trees do not produce adequate extension growth. Large scale replanting is therefore needed.

Delicious group of cultivars constitute the major share (about 83% in H.P.) of apple production in the country. These cultivars are self unfruitful and need cross pollination to ensure good fruit set. Interplanting pollinizer cultivars (Golden Delicious, Jonathan, Red Gold, Lord Lambourne etc.) in the proportion of 25 to 33 percent is necessary for good fruit set, and choice of wrong pollinizers and their inadequacy in number often result to low productivity.

In many countries, Delicious group has been replaced or is in the process of replacement with more promising cultivars. The need for injecting new blood into the apple industry through spread of new cultivars (spur types, colour mutants, strains of Gala, Red Fuji; scab resistant cultivars, bud sport selections of Royal Delicious, and some of the promising hybrids) is urgently felt. Some of the spur type and coloured mutants are already popular with farmers and high density planting has also caught the imagination of developmental departments and agencies both in H.P. and J&K. The research system has already identified Early, Mid and Late cultivars for different agro-climatic regions. The low chilling cultivars of stone fruits have also covered large tracts of the subtropical plains of Punjab, U.P and H.P. For the hills, promising cultivars identified need further spread.

Generally, apple is grown in marginal land and fertilizers are not applied according to the requirements of the trees. The water and fertilizer use efficiency is generally poor. Also, spring frost and hailstorms are adverse weather parameters leading to low productivity. Research results have shown that through proper orchard management practices (soil and water conservation and fertilizer application) the fruit yield can be doubled in the existing orchards. The adoption of improved production technology developed by the research system can bring visible and perceptible changes in the temperate fruit industry in India. Technologies like use of clonal rootstocks, introduction of renewal pruning techniques and micro nutrient applications have not been transferred and adopted at a satisfactory level.

Apple scab disease has been the major plant protection problem in apple in J&K and H.P, whilst U.P hills are comparatively free from the disease. Apple scab forecasting system developed and the chemical control schedule prescribed have helped in reducing loss due to apple scab to a considerable extent. Apple growers are adopting the prescribed schedule of chemical sprays to control the disease. For checking entries of diseased material in the free areas of U.P. and North-Eastern Hills, strict quarantine and selection of elite disease-free mother plants are very essential. Often it is not followed strictly. Some of the virus diseases have also been reported in apple for which biological and serological indexing/detection procedures have been developed. Limited quantity of virus-free budwood is also being supplied. Extreme care is now required to be taken to multiply quality planting material (in apple alone approximately 2 million plants/year) for establishing new plantations.

Most of the orchardists still sell their crop at flowering to contractors as there is no well organized marketing system. Transportation in the hills itself is problematic. Post-harvest management problems originating from poor harvesting (strip picking) and improper packing system (non CFB boxes) and lack of proper pre-cooling and cold storage facilities result in huge (25-30%) loss of fruits. Capacity of the processing sector is also inadequate. Product diversification, value addition and market infrastructure development would require very substantial investment. The existing processing units are quite old and they require modernization for which substantial investment is required. CA storage trials have shown good promise. Its extension in larger growing areas is needed. Technology for storage of apple is now known, as a result of which apple is now available throughout the year.

Government Policies and Plans for Research and Development

Research on pome and stone fruits is conducted mainly by three State Agricultural Universities, namely: a) Sher-e-Kashmir University for Agriculture Science & Technology, J&K; b) Y.S. Parmar University for Horticulture and Forestry, H.P.; and c) G.B. Pant University for Agriculture & Technology, U.P. A good number of research stations of these Universities located in major pome and stone fruit growing belts are engaged in temperate fruit research. The Indian Council of Agricultural Research (ICAR) through All India Coordinated Research Projects on Fruits and Post-harvest Technology and another Network project on Apple Scab disease has further strengthened the research activities in deciduous fruits. A few long established temperate fruit research stations namely at Shalimar Bagh in J&K, at Mashoobra in H.P. and at Chaubatia in U.P. hills have made commendable progress in temperate fruits. Both State Governments and the ICAR provide financial support to strengthen fruit research in these Universities. A large number of ad-hoc research projects on pome and stone fruits funded by ICAR are also generating good information on these crops.

During the 8th Government Developmental Plan (1992-1997) the ICAR established a Central Institute for Temperate Horticulture (CITH) with its headquarters at Srinagar in the J&K State, with a regional station at Mukteshwar in the U.P. Hills. Both these research stations will work exclusively on temperate fruit crops. This new Institute will receive major support during the 9th Plan period (1997-2001). In the North-Eastern Hills region, the ICAR Research Complex for NEH Region with its headquarters in Meghalaya State and regional units in each of the 5 other States are engaged in fruit research.

On the developmental side the State governments are engaged in nursery production of quality planting material. For example, in H.P. alone currently there are 600 nurseries in private and public sector producing and distributing more than 0.8 million plants of apple alone, every year. There are ambitious programs in all the States to further expand/replant with new improved cultivars. Apple scab disease control and post-harvest processing sectors are getting focused attention in Government developmental plans. The Directorate of Marketing and Inspection of the Government of India has framed grade standards for apple, plum and William pear. The organizations like National Horticultural Board (NHB), National Cooperative Development Corporation (NCDC), Agricultural and Processed Food Products Export Development Authority (APEDA) etc., are providing incentives to traders and exporters to improve their infrastructural facilities like grading and packaging centres, refrigerated transport, setting up of pre-cooling, cold storage, auction platforms etc. The NCDC is undertaking procurement and marketing of apple on a limited scale. The NHB has set up a market information service for the benefit of growers.

Conclusion

Deciduous fruits, covering pome and stone fruits contribute significantly to the horticulture economy of India. Apple production dominates the scene and systematic cultivation and marketing of apple can change the rural economy in the hills of North-Western India. New vision and concerted efforts are required for change in variety mix, supply of quality planting material from elite clones on indexed clonal rootstocks. High density planting, water management including micro-irrigation, integrated plant nutrient management and IPM strategy for plant protection are some of the areas which need greater R&D focus. Adoption of post-harvest management practices and infrastructure development for grading, packaging, pre-cooling and storage of the produce needs focused developmental attention. Value addition and export promotion, particularly of apple are drawing due attention of the developmental agencies in India.

2

Propagation Techniques for Tropical Fruits

The propagation of plants is of great importance in perennial/vegetatively propagated crops. When new kind of plants have to be conserved or propagated or exploited, we need to develop knowledge and technique to propagate them. Invention of propagation structures and discovery of root inducing chemicals, in addition to grafting and budding methods, have revolutionized the propagation and nursery procedures. An appropriate propagation technology can be selected for each kind of plant based on its growth, physiology, flowering and phenology.

The propagation method can be primarily designed based on the mode of natural perpetuation like species with polyembryony can be propagated through seeds. Further, the general technique standardized for the species can be applied also to other species within the same genus with suitable modifications. This could be well explained within the family Anacardiaceae where mango and cashew both can be successfully multiplied by epicotyl grafting and soft wood grafting. This paper gives details on the propagation methods followed in some of the major and minor tropical fruits of the Asia-Pacific region, such as mango, citrus, rambutan, jackfruit, litchi, durian and mangosteen.

Propagation Techniques

Mango: Mango is a cross pollinated and highly haterozygous fruit crop. Therefore, propagation by seeds (called stones in mango) lead to variability in the progeny and is a limitation for commercial orcharding, though desirable for breeding new varieties. Hence, vegetative methods of propagation are adopted for getting true-to-type plants. Seed propagation is still the chief method of multiplication of rootstocks.

Propagation of Rootstocks

Seed Propagation: Rootstock plants are raised from stones. Most of the commercial varieties in India are monoembyronic but some varieties from South India are polyembryonic which give true-to-type seedlings from nucellar embryos (Sachar and Chopra 1957). Stones collected indiscriminately give poor germination but provide a lot of variation. Bolder stones give seedlings with more vigour irrespective of whether these were collected from grafted or seedling trees. The viability of mango stones is lost very quickly. Freshly collected stones from canning units give high germination but when these seeds are stocked or kept in sun for a large period, they lose germinability.

The germination of mango stones is 80% when sown within a month of extraction but none thereafter. The viability of stones could be extended by storing them in charcoal powder (Teaotia and Singh 1971) and polythene bag (Singh and Kaisuwan 1971). Patil et al. (1986) reported 40% germination after 90 days of storage in polythene bags with charcoal. Although refrigerated storage had adverse effect on germination, seeds were successfully preserved for 12 months without any decline in viability and vigour after treatment with 1% hydroxyquinoline sulphate and storage in polythene bags of 100 guage at 15° C (Doijode 1989).

The selected stones are spread on raised nursery beds and covered with leaf mould or FYM for better germination. Germination starts within 12 - 15 days and continues up to 3 weeks. The germination is adversely affected if the stones are sown too deep or in pots or polybags. Seedlings are transplanted to pots or polybags as soon as they turn green and these are nursed and prepared for grafting. When polyembryonic seeds are sown, nucellar seedlings are separated within a month and planted separately.

Vegetative Propagation: Propagation of rootstocks by vegetative methods is advocated to eliminate variation. These methods are not commonly used probably due to absence of standard rootstock type. (1) Cutting: The hardwood leafy cuttings are used for rooting. Cutting from younger trees rooted better than those from old trees (Mukherjee et al. 1966). Rooting as well as subsequent survival has been improved by using juvenile shoots, etiolation and ringing of shoots and application of IBA in lanolin paste on the ringed portion (Bid and Mukherjee 1972). A medium containing peat moss and sand (1:1) proved better and such cuttings have to be kept under mist for rooting.

Use of bottom heat to provide a temperature of 30 to 20°C (Reddy and Majumdar 1975) for Deshehari cuttings dipped in 5000 ppm IBA and planted in sphagnum moss and sand mixture resulted in 97 per cent rooting compared to only 15 per cent without bottom heat. Beneficial effect of bottom heat was observed by Rajan and Ram (1983), Sadhu and Bose (1986), and Reddy and Singh (1987). The synergism of phenols and flavonoids with IBA for root regeneration in cuttings has been demonstrated by Reddy and Majumdar (1975). Preplanting treatment with phenolic compounds such as hydroxybenzoic acid, coumaric acid and ferulic acid generally promoted auxin-induced rooting in cuttings (Sadhu et al. 1978).

Air Layering: Etiolation of shoots before layering resulted in more fibrous roots than without etiolation and application of IBA + NAA improved rooting (Bid and Mukherjee 1969). The success of air layers was markedly improved by use of growth regulators like 5000 to 20 000 ppm NAA and IBA in lanolin paste (Chhonkar and Singh 1972).

Stooling: Majumdar and Mukherjee (1961) described modified form of stool layering, wherein after about two years of growth, the mother plant is headed back at 10-12 cm above ground to induce emergence of several shoots below the cut end. Vigorous shoots are ringed and IBA (5000 ppm) in lanolin is applied on the ringed portion about a week before earthing up. Rooting takes place in 4 to 6 weeks and these are separated from the mother plant. The same mother plant can be utilized for about ten years to obtain genetically uniform rooted plants of a given clone. Clonal propagation of mango rootstocks by stooling has also been reported by Singh and Srivastava (1982). This is the easiest and fastest method for vegetative propagation to multiply mango rootstocks.

Propagation of Scion Plants

Mango plants in nurseries are raised by vegetative methods like budding and grafting.

Budding : Various budding methods, namely, patch forkert, shield, chip and T-budding are successful in mango. However, the success varies depending on geographical location and season. Since the buddings are likely to be damaged during transport, this method is not followed commercially.

Grafting : Inarching or approach grafting, veneer grafting, wedge or whip grafting, epicotyl grafting or stone grafting and softwood grafting are the most extensively used methods.

(1) *Inarching:* This is the oldest and still the chief technique employed in many mango growing regions. Although, it gives very high success, this method is costlier, cumbersome and requires larger bud wood and a long time to produce grafts.

(2) *Veneer Grafting:* In this method, mature scion of 3-4 months is grafted on 1 to 2 years old stock. Pre-curing the scion shoot by defoliation atleast 7 to 10 days prior to grafting is essential. However, in the coastal areas of Maharashtra, prior defoliation is not required (Gunjate et al. 1976). The period of veneer grafting differs from place to place owing to climatic variations. This technique is suitable for in situ grafting and for top working of seedling trees.

(3) *Wedge Grafting:* Mature procured scion shoot of 3 to 4 months old is wedge grafted on 1 to 2 year old stock. The success is high during the humid conditions and 90 to 100% success is obtained under greenhouse and mist chamber. This method is also used for in situ grafting and top working.

(4) *Epicotyl Grafting or Stone Grafting:* In this method, mango seedling of less than two weeks is wedge grafted with mature scion. For higher success and better growth, use of two stocks per scion is recommended. This is the commercial propagation method in Konkan region of Maharashtra and does not require pretreatment of scion (Gunjate et al. 1982).

(5) *Softwood Grafting:* In this technique, grafting is done with mature, precured scion on the emerging soft copper red shoot as described by Amin (1974). This technique was successful in establishing mango orchards in situ in drier tracts where mortality of nursery raised grafts is very heavy (Amin 1978).

Citrus

The sexual as well as asexual (vegetative) methods of propagation are used to raise plants of different species in citrus. The seedlings of acid lime and mandarins are still planted in southern and northeastern region of India.

Seed Propagation : The number of seeds per fruit varies between different citrus species and cultivars. Seeds obtained from healthy, virus-free old trees which have a pedigree performance of producing vigorous, uniform seedlings should be used.

Polyembryony : Seeds of most citrus are polyembryonic and thus nucellar seedlings are used both for raising uniform rootstocks as well as for direct planting, especially in acid lime and mandarins (Singh and Singh 1955). This also helps to raise healthy plants as citrus viruses are not transmitted through seed.

Gopalakrishna and Kunte (1958) reported varietal differences in the degree of polyembryony in citrus. Motial (1963) observed that the number of embryos per seed was 4.7 in Jambhiri, 4.45 in Italian lemon and 4.25 in Galgal with a range of 1 to 11 in Citrus species. Galgal and Sadaphal had 90 and 95 per cent polyembryonic seeds respectively. Singh (1965) opined that constricted surface of seeds in Kama Khatta, Citrus jambhiri, C. aurantium and Poncirus trifoliata was associated with polyembryonic nature, except in P. trifoliata with polyembryony varying from 35.5 to 54.2 per cent. In citrus, deformed or uneven seeds were found to be usually polyembryonic (Singh 1971). Prasad and Ravishankar (1983) claimed 100 per cent polyembryony in Rusk Citrange, Trifoliate Orange and Swingle Citrumelo (93.3%), Troyer (86.6%), Carrizo (73.3%), but there was no relationship between the percentage of polyembryony and seed morphology.

Viability : Since Citrus seeds generally have no dormancy and if they are allowed to dry and the cotyledons separate, the seeds fail to germinate. Therefore, it is essential to sow the seeds immediately after extraction. The viability during storage varies depending upon species and storage conditions.

Quick evaluation of seed viability could be done by tetrazolium test soaking the seeds in 1 % triphenyl tetrazolium bromide solution which strains the living tissue.

Seed storage of Kagzi lime and Mosambi in charcoal powder at 50 per cent RH at room temperature was better than sealed polythene bags. Kumar and Majumdar (1973) reported that in C. aurantifolia, the seed germination was highest in freshly extracted seeds and declined as the storage period extended. Krishna and Shankar (1974) observed that in C. karna and Rusk Citrange, seeds viability was 100 per cent after 150 days storage in polythene bags with calcium chloride at 8°C, whereas in case of C. jambhiri and C. limonia, it was 84 and 80 per cent respectively.

Raising Seedlings : Citrus seeds are to be sown on raised and well manured beds at 2.5 cm apart in rows spaced at 15 cm. In commercial nurseries, sowing is done during mid August to February. Singh et al. (1970) observed that alkathene cover improved seedling growth of C. jambhiri, C. pseudolimon, C. limonia, C. megaloxycarpa and P. trifoliata over those in open. Many workers reported the usefulness of soaking the Citrus seeds in GA from 200 to 600 ppm to improve the rate of germination.

Vegetative Propagation

Cutting : Many species of Citrus can be successfully raised from cuttings and it is a very useful method of propagation especially when a species is desired to be colonally propagated on its own root system. However, there is wide variation in the rooting ability of different Citrus species. While limon and citron are easy to root, mandarin oranges, sour orange and grapefruit root with considerable difficulty (Sadhu 1986). Synthetic auxins (IBA, NAA and IAA) and phenolic compounds help in rooting of cuttings (Dhatt and Zora Singh 1993). However, age and physiological status of mother plant, type of wood, time of planting and media composition for planting of cuttings determine the extent of success.

Air-layering : Air-layering is fairly common and is used to propagate pummelo, lime and sweet lime. Singh (1955) achieved appreciable improvement in the rooting of air-layers of sweet lime, grapefruit and sour lime by treating the cut surface with auxins (NAA + AAA both at 1 per cent). Shankar (1965) noted that certain rootstocks like C. karna, C. jambhiri, C. pennivesiculata, C. assamensis and C. megalocarpa can easily be raised from the air-layers even without treatment of growth regulators.

Budding : Budding is the most common and widely used method of vegetative propagation of Citrus. T-budding is probably the best method and is successful with many species of Citrus. Naik (1939) exhibited that bud take in Citrus spp. increased when wood was not removed from the bud.

Stevenson (1947) reported that 'T'-budding gave 87% success in Citrus species. In Punjab, the bud take of Kinnow on C. jambhiri was 98% with better subsequent growth when round buds without wood were used than by using angular buds with or without wood (Sharma and Sharma 1986). The time of budding affects bud take and subsequent growth of-plants. Experiments conducted at different places in India have shown that February and September-October are the best periods for budding in most Citrus species.

Now polythene strips are commonly used as wrapping material instead of jute twine. Srivastava and Arya (1969) observed, while budding mosambi on C. karna, that 200 guage polythene strip was the best wrapping material. Delayed lopping of the rootstock till the bud has grown about 4-5 cm is usually practised by nurserymen.

Rambutan

The rambutan is one of the best known fruits of South-East Asia and is especially important in Thailand, Malaysia and Indonesia, but extends to the east as far as the Philippines.

Seed Propagation: Propagation by seed has been the traditional practice and large number of trees now in production throughout South-East Asia originated from seed. This is not a desirable method since plants from seeds vary and many seedlings produce poor fruits. In addition, seedlings vary in sex. Male trees are practically useless, and female trees, unless well pollinated, do not bear well. Populations of seedlings of rambutan have yielded up to 67% males. Propagation from seed should only be used for the development of new varieties.

Viability: Seeds, when removed from the fruits, have a short life time and are recalcitrant (Chin 1975; Teng 1977). It is desirable to maintain them slightly moist and to plant them in 2 or 3 days. If the seeds are sown after 3 to 4 weeks, seeds usually do not germinate.

Raising Seedling

Seeds can be sown in seed beds or in individual containers. The latter method is preferred as it eliminates the risk of transplanting and shows better establishment in the field.

Vegetative Propagation: Vegetative propagation is indispensable for production of trees of high quality. These trees tend to come into fruiting at a younger age than seedlings. To preserve good, hermaphroditic varieties, vegetative propagation is necessary (Valmayor et al. 1961).

Air-layering: This is frequently used by home gardeners and small scale producers, but success rates are generally variable.

Grafting: Methods used include various forms of approach grafting and cleft grafting, which is particularly used in the Philippines. Top working of seedling or staminate trees and grafting with improved cultivars has been described.

Budding: In most of the South-East Asian countries, seedling stocks are budded by the modified Forkert method. Various other budding methods have been used such as shield or T- budding and patch budding with limited success.

Jackfruit

Seed propagation is an old method of multiplication which is planted in situ or raised in pots. Now, vegetative propagation is the commercial method of jackfruit multiplication.

Seed Propagation: This method is mainly used for the production of rootstocks. Soaking of seeds in NAA 25 ppm for 24 hours resulted in highest germination percentage as well as good seedling growth (Bose 1986). Shanmugavelu (1971) recorded 100 per cent germination by treating the seeds with GA3 at various concentrations upto 500 ppm compared with 80% in the untreated control.

Litchi

Seed Propagation: This method is not desirable since the trees raised from seeds have long juvenile period, fail to produce true-to-type plants and often produce fruits of inferior quality. Propagation by seed is used mainly for rootstock propagation and breeding. The seeds of litchi have a very short viability, usually less than 5 days.

Vegetative Propagation

Cutting : Propagation by cuttings is possible, but the plants produced are less vigorous and this method is not generally recommended.

Air-layering : The main commercial method of propagation is by air-layering and rates of success may be as high as 95% under ideal conditions. Air-layered trees normally produce commercial crops after 3-6 years (Menzel 1991).

Grafting : Due to the limited and unpredictable activity of the cambium, which is related to the onset of leaf flushing, grafting is not always very successful. There is also possibility that some scion/stock combinations may be incompatible. However, approach grafting is generally successful. Side grafting has been reported to be successful up to 80% in Israel (Kadman and Slor 1974).

Durian

Seed Propagation : Seeds germinate within three days after sowing and there is no dormancy. Percentage of germination reduces drastically when seed moisture level depletes beyond 22%. Seedlings take longer time (8-15 years) for fruiting.

Vegetative Propagation : Budding, grafting, air-layering or even cuttings are the vegetative propagation methods followed. However, whip grafting or patch budding are the most commercially used propagation methods. Better results of grafting/ budding are obtained if rootstock of same species is used. Grafted or budded plants come to fruiting in less than 4-5 years.

Mangosteen

Seed Propagation : The best known practice for the propagation of mangosteen is by seeds and when fresh seeds are sown, they germinate in 10 to 15 days. Since the seeds are of asexual origin, they produce trees identical to the mother. If the seed is dried or kept outside the fruit for several days before planting, germination is drastically reduced. However, the seeds can be maintained for 3 to 4 weeks within the fruit. The best way to carry seeds is by transporting the entire fruit.

The size of the seeds is highly variable. The difference in seed weight brings variation not only in germination but also in later growth and survival. The ability to germinate and grow successfully is related to the amount of food stored in the seed. Therefore, it is recommended to establish new plantings only from the larger seeds, those that weigh 1 g or more.

Vegetative Propagation : Many attempts have been made to develop stronger, more rapidly growing mangoseen trees with a short juvenile phase through cutting, air-layering, Forkert budding, approach grafting and cleft grafting but the results were not promising (Campbell 1966). All the propagation methods tried did not show any real advantage over the seed propagation method.

3

Tissue Culture Strategies of Tropical Fruit Crops

A large number of tropical and subtropical fruit crop species have highly recalcitrant seeds often exhibiting viability of just a few weeks. Several vegetatively propagated tropical fruits including woody perennials need to be conserved with their genetic fidelity intact. Maintenance and preservation of germplasm of fruits in field genebanks is difficult, expensive and labour-intensive. Besides, the conection may also be vulnerable to natural perturbations.

Although conservation of seeds of economically important crops has been practised since times immemorial, it suffers from severe limitations like low seed viability and heterozygosity. Therefore, a need was felt to develop alternative methods, viz., in vitro preservation (Kartha 1981). New techniques help to conserve germplasm disease-free, but also involve lower labour costs and requirement for technical personnel besides limiting disease-transfer (Withers and Engelmann 1990).

Role of Plant Tissue Culture

The term plant tissue culture denotes genetically all cell, tissue and organ cultures, and employs the basic dogma totipotency. The technique essentially involves separation of a cell/tissue/organ from the donor plant under aseptic conditions and growing it on a synthetic medium in a suitable container under controlled environment. Among other applications, tissue culture has been successfully dovetailed to conservation and exchange of germplasm in horticulture.

In the past decades, in vitro techniques have been extensively developed for over 1000 different species. Tissue culture systems allow for axenic propagation of plant material with high multiplication rates. Virus-free plants can be obtained through meristem culture in combination with thermotherapy. The miniaturization of explants reduces space requirements and consequently labour costs for maintenance of germplasm collections. It is now mandatory to exchange coconut germplasm across countries using embryo cultures because of the obvious advantages. However, it may be noted that mere transfer to in vitro culture does not automatically confer the disease-free status.

Specific pathogenicity tests need to be carried for certification of the in vitro plant material as being disease-free.

Methods of in Vitro Conservation

Prolonged maintenance of in vitro material also provides an effective system for establishing both active and base germplasm collections particularly for the conservation and maintenance of those plant species that cannot be stored as true seed and are amenable to micropropagation technology (Fay 1994). The properties required of a successful storage system are the ability to:

- Reduce the growth rate of vitroplants, resulting in extension of intervening subculture frequencies.
- Retention of vitroplant viability during storage at maximum simultaneously maintaining genetic stability at minimum.
- Maintain the full developmental and functional potential of stored material when it is returned to physiological temperature.
- Make significant savings in labour input, materials and commitment to specialized growing facilities.

There are three types of in vitro storage for conservation:

1. Normal in vitro culture propagation.
2. Reduction in growth rate.
3. Suspending the growth.

Therefore, in vitro methods are different from either the in situ conservation in natural habitat or ex situ conservation in field conditions. Tissue cultures may be conserved at either relatively low temperature of 20 to 15°C (slow growth) or cryopreserved in liquid nitrogen at -196°C (no growth). Accordingly, In Vitro Active Gene Banks (IVAG) and In Vitro Base Gene Banks (IVBG) may be established, respectively.

Short to Medium term Storage for IVAG

The basic principle underlying this form of storage is the induction of reduced vegetative growth of the stored material by inducing osmotic stress, limiting the availability of carbohydrate to sub-optimal levels, low temperature/dark maintenance of cultures or incorporating growth retardants in the culture medium. These methods may be used singly or in combination.

Long-term Storage or Cryopreservation for IVBG

In this method, cultures are suspended at very low temperatures (-196°C) in liquid nitrogen to facilitate the arrest of mitotic and metabolic activities. The procedure guarantees long-term preservation of germplasm in genetically unaltered state. The chief drawback, however, is that the viability of very few biological materials can remain unaffected when they are frozen at very low temperatures, Cryoprotectants like dimethyl sulphoxide (DMSO) are used. Cryogenic strategies for survival of cultured plant cells are detailed by Kobayashi et al. (1991) and Sakai (1993).

Effect of Cryoprotectant on the Survival of Nuclear Cells of Navel Orange Cooled to -196°C by Slow Freezing Method

Cryoprotectant	*Survival (%)*
Control (No Additives)	30 ± 1.8
5% DMSO + 0.7M Sorbitol	30 ± 1.6
5% DMSO + 0.7M Glucose	30 ± 1.2
5% DMSO + 0.7M Sucrose	51 ± 1.2
5% DMSO + 0.7M Glycerol	3 ± 0.8

Nucellar cells were frozen slowly to -40° C at 0.5°C/min. prior to immersion into LN. Values are means of triplications ± SE (Kobayashi et al. 1990).

Choice of Plant Culture System

The basic requirement for in vitro conservation is the availability of highly reproducible protocols for either micropropagation or regeneration as the case may be.

There has been a strong, though not universal, prejudice in the past in genetic conservation programmes favouring the use of shoot cultures in particular. This is presumably because of the greater stability of such cultures compared to the adventitiously propagated systems. All culture systems from protoplasts through cell suspensions, callus, somatic/zygotic embryo to shoots/nodes and entire plantlets have valid and appropriate roles.

Effect of Sucrose Concentration in Combination with 5% DMSO on the Survival Rate of Nucellar Cells Cooled to -196° C by Slow Freezing Method

Cryoprotectant	*Survival (%)*
5% DMSO + 0.5 Sucrose	21 ± 1.4
5% DMSO + 0.7M Sucrose	44 ± 1.8
5% DMSO + 1.0M Sucrose	63 ± 1.6
5% DMSO + 1.2M Sucrose	73.± 1.4
5% DMSO + 1.5M Sucrose	29 ± 3.0

Values are means of duplicates ± SE (Kobayashi et al. 1990).

Minimal Growth Storage:

Objective (Grout 1995)

- To devise an effective and reproducible protocol for minimizing the growth and development of plant materials.
- Avoid induction of abnormal development during storage.
- Preserve the viability of the stored population of in vitro cultures.
- Ensure that the protocol allows for periodic growth under standard conditions between periods of prolonged storage.

Growth at Reduced Temperature:

- Standard culture medium in suitable containers, vessels must be robust, semi-solid medium.
- Controlled temperature and reduced lighting, light intensity can be reduced by 60% from the standard growth level.

Procedures:

1. Prepare culture vessels with appropriate standard medium.
2. Establish fresh explants from active in vitro material growing at optimal temperature.
3. Characterize the cultures.
4. Incubate the cultures to be stored at their optimal temperature.
5. Transfer the cultures to the storage temperature (4-10°C).
6. Establish at maintenance level light regime with 16-h photoperiod (at 500-1000 lux being widely accepted).
7. Inspect cultures weekly, until satisfactory conditions have been confirmed.
8. At a preselected point, the cultures should be taken from the storage conditions, transferred to the growth medium and returned to optimal growth conditions.

The Addition of Growth Regulators: Materials: Abscisic acid, CCC (Cyclocel), etc.

The Addition of Osmotic Regulators

Material: Mannitol or sorbitol.

Combined Treatments : Combination of the treatments outlined above may be of benefit.

Cryopreservation of excised meristems, shoot tips and somatic embryos (Grout 1995)

This area has not been extensively researched. There are trends and protocols with promise that have been successful for a number of species. The preferred option for extended storage of meristems is Cryopreservation and yet in this respect, the excised meristem poses particular problems due to its mass and multicellular diversity.

Objectives:

- Preservation of optimum viability of the in vitro materials when recovered from liquid nitrogen.
- To ensure that plant regeneration from viable explants following Cryopreservation is by outgrowth of original frozen thawed meristems.
- To provide the maximum genetic stability for stored explants meristem/ shoot- tips Cryopreservation by slow cooling.

Whilst there is no single procedure in this category that has universal application, there is a general approach that appears to be successful. This is:

- Pregrowth of donor plants or in vitro cultures at low temperature to bring about cellular changes associated with acclimation to low temperature.
- Pre-treatment of the excised meristem.
- Treatment with cryoprotectant.
- Cooling rate of 1°C per minute or less to intermediate temperature (typically - 40°C) followed by plunging into liquid nitrogen.
- Thawing by immersion in water at 40°C.

Procedures:

1. If the donor source of the meristem/shoot tip explants is likely to acclimate to cold, then a suitable hardening regime is to be devised When the donor material is an in vitro plantlet, a regime of temperature of day (22°C) and cool night (-1 to 4°C) may be applied for a period of 7-28 days.
2. Following aseptic excision of the meristem or shoot tip, a pre-treatment before cryoprotectant exposure and freezing is generally beneficial.
3. A suitable cryoprotectant treatment may be chosen.
4. Cooling is controlled from 0°C to an intermediate temperatures of -35 or - 40°C, at a rate between 0.1 and 1.0°C per min, with rates between 0.5 to 1.0°C per minute being preferred.
5. During storage, the level of liquid nitrogen storage vessels be regularly monitored and LN replenished.
6. Vials are thawed by immersing them individually in sterile distilled water warmed to 40°C.
7. Once thawed, the tissue is transferred to the appropriate regrowth medium.
8. There is no viability test that can be employed to assess the success of Cryopreservation of meristem/shoot tips immediately after thawing.

By rapid cooling also meristem/shoot tip can be cryopreserved, somatic embryo, can be cryopreserved by dry freezing/rapid cooling/desiccation and with/without alginate encapsulation (Grout 1995).

In short, cryopreservation techniques are of two types:

(i) Protocols/procedures having a high technology input, such as use of a programmable freezer, or methanol cryobath, etc., which enables slow cooling at variable rates upto -30 or -40°C, followed by rapid plunging in LN.

(ii) Vitrification based procedures have, in comparison to that explained in (i), low technology input, with practical advantages. They are user friendly, operationally less complex and do not require processing of samples and slow cooling rates, thus avoiding use of sophisticated equipment like programmable freezer, methanol cryobath, etc.

In Vitro Germplasm Exchange

The importance of national and international germplasm exchange in vitro can hardly be over-emphasized considering the tremendous advantages it offers over conventional methods. Besides occupying far less cargo space, it also affords exchange of axenic plant material free of undesirable contaminants.

This is of special significance in plant movement, quarantine and to breeders who can import a large number of clones of choice, multiply them and distribute them among user groups. In vitro exchange is, therefore, catching on fast. The principles are applicable to perennial fruits. It is important to use suitable, impact resistant and well sealed culture containers. The packaging should provide adequate thermal insulation and protection against rough handling. The culture medium should be of a formation that includes a higher than usual concentration of gelling agent. The most rapid available shipping method should be preferred.

Conclusions

A steady and progressive genetic erosion has resulted in a marked decrease in plant diversity. A wide genetic base helps to contain biotic and abiotic stresses that destroy entire crop stands. Developing countries, particularly those in the tropics, are still the repositories of vast biodiversity in their agro-ecosystem and act directly or indirectly as reservoirs and suppliers of germplasm.

Therefore, there is an urgent need to establish genestocks which will provide recruiting grounds for gene-hunters specially for crop varieties that run the risk of elimination by intensive agriculture. An adequate genetic resource conservation programme is to genetic engineering as a library is to knowledge. In this context, the tropical and subtropical fruit crop species stand to gain immensely from in vitro techniques of conservation and exchange of germplasm. Although many genebanks exist worldwide today, only about 30 countries provide secure long-term storage because there is little provision for long-term sustainable management of genebanks. Therefore, an integrated approach is both prudent and practical. Success stories include tropical species such as cacao, coconut, Citrus spp., coffee, etc. Even so, in vitro conservation in tropical fruit crops has a long way to go and avenues for research in this area are wide open.

Pollen Storage in Tropical Fruits

Genetic conservation through pollen storage is desirable for a variety of horticultural plant species, since pollen is known to transmit important genetically heritable characters. Pollen is a product of genetic recombination and can provide a reliable source of nuclear genetic diversity at the haploid stage. Although genetic conservation through pollen storage does not accomplish the whole genome conservation, a plant breeder involved in genetic enhancement of a given horticultural crop could have access to a facility called Pollen Cryobank, from where he can draw pollen parents of his choice in the process of breeding a new cultivar.

Pollen genepools can provide sources of resistance for exchange with other breeders (Poulos 1994). Exotic nuclear genetic diversity can be easily received and exchanged through pollen, thereby eliminating the need to go through a long juvenile phase, common in most fruit trees to produce pollen for hybridization at a desired location. Thus, stored pollen can be used to improve breeding efficiency. Plant breeding strategies could derive the benefit of pollen selection when samples having increased longevity are indexed as viable and fertile for carrying out breeding operations throughout the year (Hanna and Towill 1995).

Fruit tree pollen is generally required to be stored for controlled crossings, either to achieve a desired breeding objective, or to overcome a constraint involved in commercial fruit production. For example, in Citrus rootstock breeding programme, interspecific or even intergeneric crosses are attempted to introgress genetic traits into the desirable cultivated plants from pollen parents known for providing resistance to biotic stresses. Citrus and its allied species flower asynchronously and successful crosses are seldom accomplished with fresh pollen due to non-availability of the seed parent when the pollen parent is in full bloom. Pollen storage has come to the rescue, where stored pollen indexed as viable can be used in crossing with the desired female clone so as to accomplish the breeding objective.

Successful transfer of trifoliate leaf character to the F1 zygotic hybrids of an intergeneric cross involving Poncirus trifoliata as the pollen parent and Citrus limonia as seed parent, using one year cryopreserved pollen (Rajasekharan et al. 1995) was accomplished. On the same lines, pollen of mango cultivars were successfully cryopreserved for two years. Pollen fertility was retained for five years as tested by intervarietal crosses.

Genepool conservation at the haploid stage can, therefore, be effectively accomplished through pollen which can provide a rich source of nuclear genetic diversity. A major emphasis on research needs include pollen storage (Arora and Rao 1995) for citrus, mango, rambutan, jackfruit, durian and litchi. Pollen cryopreservation research has been recently recognized by IPGRI (IPGRI 1995). As one of the genepool components in an integrated PGR conservation programme, pollen can serve as an alternative or additional ex situ conservation method. Alexander and Ganeshan (1993) have extensively reviewed pollen storage research in fruit crops. Grout and Roberts (1995) have elaborately described the methodology involved in pollen cryopreservation. Hoekstra (1995) has assessed the merits and demerits of pollen storage for genetic resources conservation. The present lecture will give an insight on research progress made at IIHR with respect to citrus and mango pollen storage.

Citrus

Pollen Collection : The species handled for pollen collection include Citrus limon, Citrus aurantifolia, Citrus sinensis and Poncirus trifoliata.

Inventory : Petri dishes, butter paper, forceps, muslin cloth fixed to a 10 cm. cylindrical ring with the help of a firm rubber band, clean razor blade, etc.

Procedure : Pollen collecting are usually made on a bright sunny day between 8 - 10 AM, staminate flowers are harvested at peak anthesis and the dehiscing anthers are gently caressed over a muslin cloth sieve. Anthers are carefully removed from pistillate flowers and pollen is extracted over a muslin cloth sieve. Thus, pure pollen is extracted in clean petri plates or butter paper.

Precautions : The following points should be noted:

1. Quality of pollen collected depends on the correct stage of anthesis/anther dehiscence.
2. Pollen should be free of anther debris.
3. Do not collect pollen from infected or insect damaged flowers.
4. Decide upon bagging of inflorescence depending on insect activity.
5. Do not venture to collect pollen on a rainy day or if it had rained overnight.
6. Do not force out pollen from anthers.

Viability Testing : Pollen collections are subjected to viability indexing by germination in vitro by the hanging drop technique (Shivanna and Rangaswamy 1993). This process involves suspension of pollen grains in a drop of liquid medium consisting of a carbohydrate source, inorganic and organic compounds.

Inventory : Cavity slides, cover glasses (22 mm square), microslides, petroleum jelly, large petri plates, filter paper discs, incubator, double distilled water, needle, forceps, "Quick fix" (epoxy resin).

Medium : Prepared in deionised, double distilled water. 20% sucrose, 100 ppm H_3BO_3, 300 ppm $Ca(NO_3)24\ H_2O$, 200 ppm $Mg\ SO_4\ 7H_2O$, 100 ppm KNO_3, pH 7.3.

Procedure : Apply petroleum jelly to the edges of a new cover glass. Place this cover glass on the work table so that the edge on which petroleum jelly is applied faces the roof. Place a 25 to 50 l drop of nutrient medium on the centre of the cover glass assuming the shape of a small bubble. Add pollen grains with the help of a needle carefully filling the drop with optimum quantity of pollen. Invert a cavity slide over the cover glass with the cavity over the drop.

The petroleum jelly helps to adhere to the cover glass around the cavity, sealing the cavity from all sides. Reverse the cavity slide and place in a petri plate fixed with moistened filter paper disc cut to its exact size. Place the cavity slide over a microslide in such a way that the cover glass with the hanging drop over the cavity is well clear of its surrounding area, intact and hanging. Cover petri plate with the top lid (also fixed with moistened filter paper) and keep in an incubator at 25 ± 2°C, dark, for 18 to 20 hours.

After the prescribed duration of incubation, the hanging drop preparation is separated carefully from the cavity slide. After removing the residual petroleum jelly from the slides of the cover glass, a drop of Alexander's differential stain (Alexander 1980) is added over the stained drop and mixed slowly. Place a clean microslide over the stained drop. The drop is flattened with the cover glass sticking to the underneath of the microslide. Remove excess stain and seal the edge of the cover glass microslide interface with a thin layer of quick acting adhesive. The slide with germinated pollen is ready for observations under the microscope for qualitative and quantitative viability estimates.

Precautions : Noteworthy points are:

1. Prepare medium fresh.
2. The medium on the cover glass should be just optimal for it to hang on the cover glass when inverted.
3. Add just enough stain to the drop containing germinated pollen. If excessive stain is added, the germinated pollen drifts away outside the cover glass when the drop is flattened, leading to errors in quantitative estimation.

Storage

Inventory : Butter paper, glass vials, silica gel, gelatin capsules, polyethylene aluminium laminated pouches, head sealer, domestic refrigerator, freezer, liquid nitrogen storage, cryoflask filled with liquid nitrogen.

Procedure : Pollen samples are packed in glass vials/gelatin capsules/butter paper packets depending on the method of storage. Lemon pollen contained in glass vials are best stored at 5 or -18°C for short duration of 8 weeks (Ganeshan and

Sulladmath 1983) and medium term duration of 6 months (Ganeshan 1983) with a progressive decline in viability index by germination in vitro at weekly or monthly intervals. Cryogenic temperatures (-196°C) afford much more prolonged duration storage, beyond two years as demonstrated by Japanese workers (Kobayashi et al. 1978). Ganeshan and Alexander (1991) reported cryogenic preservation of lemon pollen for 3.5 years.

Storage of 5°C and -18°C : Storage in liquid nitrogen: Pollen samples are packed either in gelatin capsules or butter paper packets, sealed air tight in polyethylene aluminium laminated pouches and lowered into a canister of a cryoflask. The canister is capped with a perforated lid and plunged slowly in liquid nitrogen contained in the cryoflask. Frequent refilling of the cryoflask with liquid nitrogen at least once every 10 to 15 days, ensures a constant cryogenic temperature. For proper ease of operation, samples to be cryopreserved for a long duration are located at the bottom of the canister and those that are to be retrieved after a short duration, at the top of the canister.

Retrieval and Post-storage Viability Assessment : There are not many critical steps to be followed for retrieval of pollen samples stored at low or sub-zero temperatures. Pollen is removed from the glass vial and directly tested for viability as described above. Pollen samples retrieved from cryogenic temperatures have to be carefully pulled out of the canister with the help of a one foot long blunt forceps, held at ambient temperature for 10-15 minutes prior to a viability test or field pollination.

Pollen samples are viable if:

- Germination profiles in vitro are as good as fresh pollen.
- Germination time taken is not more than that of fresh pollen (pollen vigour).
- Estimated germination is in the range of 60 to 80 per cent of fresh pollen.

Field Pollinations with Stored Pollen for Fertility Assessment

Inventory : Breeder's crossing kit.

Procedure : Stored pollen are required to be tested for fertility under field conditions through controlled pollination. For this, pollen samples retrieved from storage are transported to field locations in an ice box and used in crosses with compatible seed parents in order to obtain an estimate of fertility, fruit and seed set.

Stored pollen is applied to pre-bagged emasculated perfect flowers at peak stigma receptivity with the help of a painting brush, followed by bagging to avoid contamination with stray pollen. For comparative fertility estimates, crosses with fresh pollen must be carried out as far as possible at least initially. If the pollen parent in question is not in flowering, crosses may be repeated with the fresh pollen over the same seed parent within a short period.

Observations to be recorded in field are:

- Number of crosses made with fresh and stored pollen.
- Number of fruits set to maturity.
- Number of seeds per fruit.
- Number of aborted seeds.

Seeds set through stored pollen are germinated and observation of any marker characters transmitted by pollen parent (for example, trifoliate leaf character in F1 population of crosses involving pollen of Poncirus trifoliata), would confirm that the stored pollen was fertile and has transmitted the character conserved to the next generation.

Mango

The species handled for pollen collection include Mangifera indica.

Pollen Collection : Each inflorescence consists of several thousands of staminate and perfect flowers, having a preponderance over staminate flowers. Anthesis is initiated much earlier to the complete development of the inflorescence:

Inventory : Petri dishes, butter paper, forceps, polythene covers (for bagging if necessary) needles, clean razor blade, etc.

Procedure : Pollen collection is usually synchronized with anthesis, between 9 and 11 AM. Anthers dehisce only after the flower is opened. In most Indian cultivars, staminate and/or pistillate flowers have just one or occasionally two anthers producing viable and fertile pollen. These are about 2 mm long and could be seen with the naked eye. In some cultivars, the anthers are pink, turning purple on anthesis, and in some they are dark maroon turning to weak ash-grey colour which is "pollen covered" on the dehisced anther.

Pink or dark maroon anthers are picked out from each flower using a fine pair of forceps and kept on a butter paper in a petri plate. Although it is a laborious process, considerable amount of such anthers are consolidated on a butter paper kept in a petri plate, These anthers on dehiscence turn purple or ash grey. A hand lens can reveal the presence of pollen on the dehisced anthers.

Viability Testing : Spencer and Kennard (1955), and Young (1956) reported in vitro germination of mango pollen. Germinating pollen by the hanging drop method as described for citrus or the modified cellophane method (Alexander and Ganeshan 1989) for recalcitrant pollen resulted in poor viability estimates (Rajasekharan 1996, personal communication). Efforts are under way at IIHR to design a repeatable procedure for germinating mango pollen.

Storage

Inventory: Gelatin capsules, polyethylene aluminium laminated pouches, heat sealer, cryoflask with liquid nitrogen.

Procedure : Mango pollen is required for breeding only once a year and hence long term cryogenic preservation effect was investigated. Anthers laden with pollen were enclosed in gelatin capsules, sealed in polypouches and lowered in canisters and capped as described for citrus. Complete immersion of the samples are ensured by periodical replenishment of the cryogen.

Retrieval and Post-storage Fertility Assessment

Inventory : Field pollination kit.

Procedure : Post-storage fertility of mango pollen is tested by field pollination. At peak stigma receptivity, emasculated flowers of seed parent previously enclosed in polythene bags were pollinated with cryostored pollen by gently touching the receptive stigma with the anther-pollen mixture drawn from the gelatin capsule with the help of fine brush, followed by bagging. Fruit and seed (stone) set could be observed in the next 2 to 4 weeks.

Due to a high inherent fruit drop rate (Mathews and Litz 1992), very few fruits develop among crosses involving cryopreserved pollen as observed in the case with fresh pollen crosses. Cryopreserved mango pollen induced fruit and seed set among intravarietal crosses attempted at IIHR, thereby demonstrating that mango pollen could retain fertility after long term cryopreservation. Further studies on the seed (stone) set with stored pollen is in progress.

Some theoretical and practical considerations of cryogenic preservation of pollen:

1. Techniques of cryogenic storage are new to many plant species, of which the success rate is very high for bicellular pollen material, often collected from tropical regions.
2. Survival of pollen to cryogenic storage and eliciting a high viability response can be used as a model to study the moisture kinetics involved in other plant tissues.
3. The cryogen used *i.e.,* liquid nitrogen though inert, is hazardous by way of causing frost injuries with risk for the personnel using, and those who work in the surrounding area. Training in cryogenic operations, handling procedures, safety precautions, etc., are essential.
4. Places with cryoflasks filled with nitrogen should minimise the risk of gaseous nitrogen build-up. Small volumes of liquid on expansion at ambient temperature, converts into large volumes of gaseous nitrogen, reducing the local oxygen content. This can cause drowsiness, and in extreme cases, asphyxiation.
5. Proper handling equipment, such as cryogloves, long forceps or tongs, face guards if necessary while working for long duration, is essential. It is generally

advised not to use glass vials for cryogenic storage, as they can explode due to seepage of liquid nitrogen into the vial during storage.

6. Liquid nitrogen source as far as possible, must be located in proximity and should be easily accessible.

Cryogenic Containers

- Commercially available storage containers with a long static liquid nirogen holding time and with a low evaporation rate are desirable. The canisters designed to hold vials, pouches, etc., in liquid and vapour phase are to be used.
- It is of cardinal importance to ensure transfers or retrievals in and out of the canisters as quickly as possible, so that samples held in storage at -196°C are not allowed to warm up significantly as a result of being relocated temporarily.
- The liquid nitrogen level in storage cryocans will have to be monitored regularly (at least once every week and more frequently if the containers are often opened within this period). Accordingly, the cryogen level has to be replenished to the level recommended by the manufacturer to achieve the required temperature. Keeping the containers in a room temperature controlled between 0° - 10°C reduces loss of the cryogen to a great extent, thus extending the time between fillings.

Tropical Fruit Germplasm

A considerable variety of tropical fruits are grown in Asia, in particular in South and South-East Asia. The region is a centre of diversity for several important native fruits which include mango (Mangifera indica), Citrus and rambutan (Nephelium lappaceum) (Arora 1994; Kajiura 1990). In addition, more than 200 minor, underutilized native species occur; these are collected in the wild or grown in home gardens. Some of these species have considerable potential, such as breadfruit (Artocarpus altilis) and jackfruit (Artocarpus heterophyllus), which are rich in carbohydrates, proteins, minerals and many vitamins (Negi et al. 1990). Examples of minor fruit crops with high genetic variation in the region are carambola (Averrhoa carambola), longan (Dimocarpus longan) and aonla (Phyllanthus emblica) (Arora 1994).

Just as for other crops, it is important to conserve the genetic material of the various fruits. Human beings are highly dependent on plants for survival. Plants are mainly used for food, but also as sources for building material, fibres, medicines, oils, rubber, perfumes and dyes. The relationships between human being and plants are not static. Because of constantly changing needs in human nutrition, increasing population and industrialisation and trade, a large variety of species is required to meet these needs.

In India and its neighbouring countries, various native fruits, such as aonla, beal fruit (Aegle marmelos), jackfruit, jamun (Syzygium cuminii), karonda (Carissa congesta), Kokum (Garcinia indica) and phalsa (Grewia subinaequalis) are underutilized. Some of these might be important in the near or far future, because of their therapeutic/medicinal and nutritive value as well as their excellent flavour and very attractive appearance. Consumers today are becoming increasingly conscious of the health and nutritional aspects of their food.

Underutilized fruits could play an important role in satisfying demand for nutritious, pleasantly flavoured and attractive natural food of high therapeutic value. Encouraging local people to produce these fruits can help to improve their social and economic welfare. In this way, they can also significantly contribute to the preservation of the environment by stopping uncontrolled harvesting from the wild and assisting in the retention of the various species in their native habitats where they perform best.

Conservation is very important, because many species are becoming extinct and many others are threatened and endangered. The diversity of some fruits is well collected, while for other fruits relatively little has been done yet (Arora 1994). Gaps in collections are found both between species and between regions. This is especially true for both underutilized species and wild crop relatives, where big gaps are noted. Kostermans and Bompard (1993) indicate that Mangifera blommesteinii, M. leschenaultii, M. superba and M. paludosa are in real danger of extinction. High genetic erosion has been noted for jackfruit, Citrus sp. and Litchi chinensis in a survey carried out by the International Centre for Underutilized Crops (ICUC) and IPGRI (Haq 1994).

Furthermore, great genetic diversity is required in order to have a stable ecosystem. In the case of species, which are already used by human beings as crops, it is very important to have a broad genetic base, to improve existing genotypes when necessary. Finally, the diversity of plants in different ecosystems brings a lot of pleasure and inspiration to people with cultural and/or religious significance and the potential for income generation through eco-tourism. Thus, it is important to appreciate the contribution to human welfare and environmental sustainability made by all the three levels of biodiversity: (i) ecosystems, (ii) species, and (iii) genetic diversity (IPGRI 1993).

This paper deals with in situ conservation, particularly as it relates to tropical fruit crops. Two different approaches of in situ conservation are distinguished, namely, genetic reserves and on-farm conservation. Before going into detail on both approaches, major problems and constraints for in situ conservation research are reviewed to give a better understanding on the status of in situ conservation at present. Lastly, various examples, a few on tropical fruits, are presented on how in situ conservation could be applied. These examples can provide a basis for further discussion on how to approach and apply in situ conservation.

Definitions

In situ conservation is defined as the conservation of ecosystems and natural habitats, and the maintenance and recovery of viable populations of species in their natural surroundings and, in the case of domesticated or cultivated species, in the surroundings where they have developed their distinctive properties (UNEP 1992).

Ex situ conservation, on the other hand, is defined as the conservation of components of biological diversity outside their natural habitats, for example, in genebanks (CBFD 1992). Thus, conserved species can be either wild or domesticated and habitats can be either natural ecosystems or agro-ecosystems. In practice, the situation is more complex as different intermediate types exist. For example, IPGRI divides its mandate crops into the following groups: cultivated species and their wild relatives and useful forestry species (Iwanaga 1995). Brush (1991) distinguishes the following four categories of crop germplasm: wild crop relatives, semi-domesticated (weedy) crop relatives, perennial species and landraces of ancestral crop species.

In this paper, the in situ conservation strategy is divided into two general approaches, following the working definitions proposed by Maxted et al. (1997).

1. *Genetic Reserve Conservation:* The location, management and monitoring of genetic diversity in natural wild populations within defined areas designated for active, long-term conservation.
2. *On-farm Conservation:* The sustainable management of genetic diversity of locally developed traditional crop varieties with associated wild and weedy species or forms by farmers within traditional agricultural, horticultural or agri-silvicultural cultivation systems.

In Situ Conservation

Traditional, many crops are conserved as seed in genebanks. However, conservation of tropical fruit crops often takes place in field genebanks due to the recalcitrant nature of the seeds of many of them. Major disadvantages of field genebanks, such as high maintenance costs, the limited amount of genetic variation that can be stored and vulnerability to natural and human disasters have led to efforts to develop in vitro conservation methods.

This conservation can be either under slow growth conditions or through cryopreservation (preservation at very low temperatures in liquid nitrogen). Pollen and DNA conservation are two other techniques, which are being explored and are used on a small scale. All of the above mentioned ex situ conservation techniques have their niche in conserving part of the genetic variation of fruits. However, the conservation of plant material outside its natural habitat brings disadvantages. For example, to conserve material ex situ, it is necessary to remove it from the area in which the plants occur naturally, so interrupting the evolutionary process. Risk of genetic drift increases. This disadvantage can be resolved by using an in situ conservation strategy. However, this does not mean that in situ conservation is the only or best way of conservation; rather it should be seen and used as an additional and complementary technique. In situ conservation can be a valuable tool, because it extends the conservation of a species beyond the level of the individual to the habitat or ecosystem in which the species lives.

In addition to preserving the plant material per se, the processes that create new genotypes are also conserved. In other words, evolutionary processes continue letting the plant populations continually adapt to the changing environment. Furthermore, in many case's, it is easier to conserve a viable population in situ than ex situ. This is certainly true of tree species. In situ conservation might be of great importance for: (i) recalcitrant seeded species, (ii) wild relatives, (iii) wild species, and (iv) species depending on other organisms within the ecosystem, *e.g.*, some plant species have highly specialised breeding systems, which require insect or birds for pollination. For the first group, other solutions which are often very expensive, might result in the conservation of a limited number of accessions. The second and certainly the third group often receive a low priority in collecting, which could be resolved by in situ conservation. For the last group in situ conservation is the only way.

Major Problems and Constraints for in Situ Conservation

At present, there is substantial body of information available on the development and establishment of biological reserves. However, there is limited information on the development and establishment of genetic reserves and only very limited knowledge is available on on-farm conservation. Knowledge on the genetic diversity of the target species, agro-ecology and indigenous knowledge are recognised as important elements for on-farm conservation, yet a logical framework is lacking. Iwanaga (1995) recognised and distinguished the following problems and constraints for in situ conservation of crops, their wild relatives, and useful forest and non-forest plant species:

- The absence of a formal institutional framework able to develop and support in situ conservation activities;
- Lack of linkages (both formal and informal) between different groups interested in maintaining or conserving different species and types of plant germplasm in situ.
- Inadequate knowledge of communities who maintain crop or other plant germplasm in situ and an absence of systems for supporting such communities (and frequently, a lack of recognition of their role and interests).
- Low level of awareness by formal-sector agencies of the significance of in situ conservation and the contribution that it can make to biodiversity maintenance and to provision of useful germplasm as a contribution to development.

- Limited number of trained personnel able to guide and lead the development of in situ conservation activities of crop, forestry and other plant species.
- Major limitations in the knowledge base with respect to selection of in situ conservation locations, their monitoring and management.
- A lack of effective analysis of the relationship between development and conservation of the different forms this relationship can take and the impact these will have on the maintenance of the diversity. In the next part of this presentation, existing approaches and frameworks for in situ conservation are described. However, before that, different activities that should be undertaken in order to be able to develop and establish final conservation plan are presented.

Preliminary Information Gathering for in Situ Conservation

Before planning and designing a genetic reserve or an on-farm conservation project, several issues need to be considered. Firstly, the main target and related species should be identified. The choice for these should be objective, based on logical, scientific and economic principles related to value and use of the species. Other important factors influencing the choice of a target are threats, which vary from species to species. These threats can be grouped under the headings of overexploitation, habitat destruction or fragmentation, competition from introductions, changes and intensification of land use, environmental pollution and climatic change (Maxted and Hawkes 1997).

A practical approach to select target species could be the use of the genepool concept of Harlan and de Wet (1971). This concept is based on the ease with which species hybridize with each other. Three categories of genepools are proposed:

- *Primary Genepool (GP-1):* This includes the true biological species including all cultivated, wild and weedy forms. Crossing between the various species within this genepool should not cause any problems and the derived hybrids should generally be fertile.
- *Secondary Genepool (GP-2):* This includes those biological species that can be hybridized with the main species, but often result in sterile progeny.
- *Tertiary Genepool (GP-3)*: This includes the species that only could be hybridized with the main species with difficulty, *e.g.*, use of embryo culture, doubling the chromosome number or using bridging species. Resulting hybrids are anomalous, lethal and/or sterile.

The availability of information on different genepools enables the priority setting of target species to be incorporated into the in situ conservation strategy. The general importance of a species and the number selected within a species depends on the ease with which they can be used. So this means that the highest priority should generally go to species in the primary genepool and the lowest to those species in the tertiary genepool (Chapman 1989). In formulating strategies for the conservation of any crop, it is essential to know its areas of distribution, and to understand the diversity and location of its wild relatives. This knowledge base should consist of both geographical and ecological information. Furthermore, information on the density of the different species is needed.

In order to conserve biological diversity, conservation programmes must be guided by the biology of the species or systems that one seeks to preserve (Karron 1991). The taxonomic status, geographical range, life form, breeding system and seed dispersal mechanism all have significant effects on the levels of genetic diversity maintained by a species (Hamrick et al. 1991). These factors influence the genetic diversity within and between populations of the target species.

Dicots generally have a lower genetic diversity than monocots and gymnosperms. Annuals and short-lived perennials and endemic species normally have a lower genetic diversity than long-lived perennials and wide spread species. Higher diversity is found among cross pollinating species than in self-pollinating species.

The foregoing information indicates that species with limited potential for gene flow have more differentiation between populations, while species with a high potential for gene flow have more differentiation within the population. Gene flow potential is closely correlated with type of pollination, seed dispersal mechanisms, longevity and seed size. Hamrick et al. (1991) concluded that species with high levels of genetic diversity are more likely to have populations with high genetic diversity than those species with low levels of genetic diversity.

In addition to the various technical aspects influencing the choice of target species, and location and size of genetic reserves, socioeconomic and political factors also play a role. Firstly, there are various international agreements and conventions influencing and impacting upon the conservation of plant genetic resources, among which more important are the following:

- The Convention on Biological Diversity (CBD).
- Agenda 21.
- Resolution 5/89 on Farmers Rights.
- FAO Global Plan of Action for the Conservation and Sustainable Utilization of Plant Genetic Resources for Food and Agriculture.

Article 8 of the CBD, a legally binding convention, accepted by the Conference for the Adopting of the Agreed Text of the Convention on Biological Diversity met at UNEP Headquarters in Nairobi, Kenya in 1992, deals with in situ conservation. Although this is a fairly long article, it can be summarised as follows: the contracting parties should establish a system of protected areas to conserve biological diversity. This should be done in such a way that conservation is ensured and that diversity will be used in a sustainable way. Guidelines for the selection, establishment and management of protected areas should be developed where necessary. These legally binding norms provide a broad basis for both genetic reserves and on-farm conservation.

In situ conservation is also discussed in Chapter 14.G of Agenda 21. Two major objectives of this programme are:

1. to adopt policies and strengthen or establish programmes for in situ conservation;
2. to take care of sustainable use of plant genetic resources for food and agriculture, integrated into strategies and programmes for sustainable agriculture;
3. to take appropriate measures for the fair and equitable sharing of benefits and development in plant breeding between the sources and users of PGR".

Resolution 5/89 on Farmers Rights was added in 1989 to The International Undertaking on Plant Genetic Resources. Resolution on Farmers Rights aims to (Henne 1995):

1. ensure that full benefit to farmers from the exploitation of the diversity that farmers have generated goes to the farmers themselves;
2. support the continuation of the contributions by farmers to the development and maintenance of landraces;
3. attain the purpose of the undertaking. The purpose of the undertaking according to Art. 1 is to ensure that "plant genetic resources, especially species of present or future economic and social importance, are explored, collected, conserved, evaluated, utilized and made available, for plant breeding and other scientific purposes."

Farmers Rights differ from plant breeders right (PBR), in that the objective of Farmers Rights is the conservation and sustainable use of biological resources to serve the community, rather than allowing that biological resources to become a commodity. Furthermore, these rights are a counterpart to intellectual property rights (IPR). In addition to establishing funds to support farmers in protecting and conserving genetic diversity and to inform the public about the importance of conserving these genetic resources, it also guarantees full benefit to farmers of the gains which are realised by others using the plant genetic resources developed by farmers (Henne 1995).

The GPA is a set of recommendations and activities arising from the Report on the State of the World's Plant Genetic Resources, prepared in 1996 on the basis of 150 country reports. The GPA was adopted at the FAO International Technical Conference on Plant Genetic Resources held in Leipzig, Germany in June 1996.

The GPA main objectives are (FAO 1996) to:

- ensure the conservation of PGRFA as the basis of food security.

- promote sustainable use of PGR to foster development and reduce hunger and poverty.
- promote the fair and equitable sharing of the benefits arising from the use of PGR.
- assist countries and institutions to identify priorities for action.
- strengthen existing programmes and enhance institutional capacity.

The GPA refers to both in situ and ex situ conservation. For in situ conservation, four main activities are recommended in this book of the GPA. Firstly, PGRFA should be surveyed and inventoried, in order to be able to develop rational conservation systems and national policies for the use of these resources. Secondly, more support through training, participatory research and a supporting policy environment will be provided for the management and improvement of on-farm conservation.

Thirdly, assistance will be offered to farmers in disaster situations, such as war, civil strife or natural disasters to restore agricultural systems. Lastly, the in situ conservation of wild crop relatives and wild plants for food production will be promoted through improved management of the genetic resources in genetic reserves.

National policies should deal with the national agricultural policy, the national research agenda, genetic conservation programmes, farmer consultation by official agencies, education and incentives for farmers and farming resources (Henne 1995). The 1992 National Integrated Protected Areas System Law of the Philippines is an example in which the legal instrument specifically addresses genetic conservation management.

One of the problems which can be encountered here is that such laws can actually cut across the mandated areas of responsibility of a number of different ministries and departments, such as agriculture and environmental affairs. This makes the application and enforcement of these laws very difficult.

Lastly, it is extremely important to get local communities involved, because they are socially and economically dependent on the environment where the conservation should take place, especially for on-farm conservation. However, even in the case of natural reserves, local communities are often dependent on those areas for example for the collection of food, medicinal plants and fuel wood. The collected material can be either for own use or cash income.

Thus, it is important to explain the purpose of in situ conservation to local communities, to obtain information from them on the area, and to exchange ideas about the selection of a site, using rapid appraisal and participatory appraisal techniques. In other words, indigenous knowledge should be considered as a valuable source of information. Finally, it is very important that communities feel and take responsibility for the entire project. This can make the difference between success and disaster.

Genetic Reserves

A model for how to plan and establish a genetic reserve has been developed by Maxted et al. (1997). This model divides the entire process of establishment and use of genetic reserve into three phases:

1. Reserve planning and establishment.
2. Reserve management and monitoring.
3. Reserve utilisation.

Reserve Planning and Establishment

Using ecogeographic studies, broad areas of genetic diversity are identified. For the establishment of the actual reserve, precise sites have to be defined within these broad areas. Potential sites require a detailed survey based on taxonomic diversity of target species in order to have detailed and up-to-date information on the various sites to facilitate the selection of best sites. This assessment of genetic diversity increasingly involves the use of molecular techniques, such as isozymes or DNA markers. Furthermore, the habitat requirements of the target taxa should be verified and the most suitable sites should contain robust populations of the target taxa.

The sites should hold as much genetic diversity as possible, because the minimum viable population size is often not determined in advance of conservation (Maxted et al. 1997). To prevent total loss of a target species, it is advisable to establish more than one reserve, selecting sites which complement each other. This means that the widest possible range of different ecological and geographic zones in which the target species are found, should be covered.

When considering potential sites for establishing genetic reserves, it would be advantageous to link these to other ongoing conservation interests. In the wider conservation movement, the types of protected area are fairly well understood as defined by the IUCN Commission on National Parks and Protected Areas, and also by various international networks and conventions. As far as possible, genetic reserves should be designed within the general framework set out by these agreements.

Some sort of cost-benefit analysis should be carried out for each site in order to be able to select the most valuable one. It is not easy to measure the benefits, as it is very difficult to estimate the value of a resource especially when it is not of particular interest at this moment, but which might have a value in future. This often means that these figures cannot have more than moral and scientific value. The relative costs of establishing a reserve also affect selection of the site. Obviously, if two sites are of equal genetic value, the one which will be the cheapest to establish should be selected. Due to problems in estimating benefits, the cost-benefit analysis cannot be decisive for the site selection; rather it is an additional information source.

Subsequently, the reserve site should be sustainable in this role for foreseeable future. It must be accessible to the reserve manager and potential users. When selecting a site, the national regulations and policy on plant genetic resources should be kept in mind. It is very important to involve the local community and this should be done at as early a stage as possible.

After the site for a reserve is determined, the reserve itself should be designed. Factors, such as structure, size, whether a single large or multiple smaller sites for the target species, the use of corridors, reserve shape, environmental heterogeneity and potential user communities have to be considered (Maxted et al. 1997).

According to the Man and the Biosphere Programme (UNESCO), a genetic reserve should consist of a core zone surrounded by a buffer zone and outside that, if possible, a transition zone. The core zone should represent a natural undisturbed or minimally disturbed area, representative of the ecosystem. Certain degree of management need to take place and the core species need to be monitored. The buffer zone protects the core zone from edge effects and other factors that might threaten the viability of the target populations in the core. It also allows the reserve to have a more natural boundary. In addition, the buffer zone allows restricted use of the traditionally utilized species by the local communities and it can be used for research, education, demonstration and harmonious interaction with people. The transition zone serves to shield the core zone from general areas of human exploitation.

This zone may be used for the development and research of proper techniques for eco-restoration, limited human settlement, sustainable utilisation and tourism (Hawkes et al. 1997). The size of a genetic reserve should be large enough to ensure self-perpetuation and continuing evolution of the biodiversity within it (Arora and Paroda 1991). This means that the core zone should contain at least 1000 to 5000 potentially breeding individuals to represent a population, the minimum viable population size (Hawkes et al. 1997).

The reserve design can be based either on the use of various small reserves or on one large reserve. For example, is it better to have one large reserve of 15 000 hectares or a network of five, each of 3000 hectares? A large reserve has as advantage that it might cover more habitat types than small reserves. However, the different small reserves can be placed in different environment to enable conserve all extreme ecotypes. Of course, it should be kept in mind that small reserves should be big enough to contain the minimal population size to survive.

The size and number of reserves depend on the characteristics of the target population. Large areas are better for maintaining larger species, while smaller multiple reserves might be appropriate for annual plant species that are often found in dense but restricted stands (Maxted et al. 1997). For the majority of the wild relatives of crop plants, the establishment of multiple small reserves is preferred, because they are generally widely distributed over a range of different ecosystems. The conservation value of multiple small reserves can be increased by the establishment of corridors which facilitate gene flow and migration between the component reserves.

The edge of a reserve should be kept to a minimum to avoid side effects from the outer environment, such as changes in light, temperature, wind, the incidence of fire, introduction of alien species, grazing and deleterious anthropogenic effects. For that reason, a round-shaped reserve has the lowest edge to area ratio. It should also be realised that each disturbance in the reserve, such as roads, dams and fences causes additional edge effects (Maxted et al. 1997).

Lastly, the sustainability of the reserve needs to be considered. In comparison with the maintenance costs of ex situ conservation, the maintenance costs for reserves are high. This is due to the need for active and consistent population monitoring, habitat management and site security, all of which require the commitment of substantial level of resources for a substantial period of time. If material is lost during the in situ conservation process, it could be reintroduced, if stored at another place, either in another reserve or ex situ.

It is also important for sustainability to obtain as much knowledge as possible about future plans for human development for the site selected. Legislation ensuring that once a reserve has been established, it is maintained to assist in securing the sustainability of the reserve (Maxted et al. 1997).

Reserve Management and Monitoring

To be able to manage a reserve in a proper way, a good management plan has to be established. Management plans normally contain information on conservation objectives, site biotic and abiotic dynamics, site history, public interest, factors influencing management, management prescription (what work needs to be carried out and precisely how and when to do it), ecological and genetic survey and monitoring schedule, budget and human resources (Maxted et al. 1997).

In the case of the conservation of certain target species in a reserve, information should be added on the species themselves. This information should be both general (taxonomy, phenology, habitat preference, breeding system and minimum population size), and specific, based on the site (mapping of populations, information on density within the site and interaction of species with the surrounding environment). Furthermore, the plan should contain information on what is an acceptable fluctuation in size of the populations (maximum and minimum values) and when to raise the alarm to take alternative actions to save populations.

A good management plan should also contain detailed information on the appropriate level of intervention through grazing, mowing, burning, planting and harvesting regimes (Maxted et al. 1997). When the management plan is introduced, it should be continuously combined with evaluation, revision and refinement in the light of its practical application. An example of how management can influence a population is seen in the cutting or non-cutting of branches of Myrica trees in Japan. In the protected forest zone along the coastal cliff at Yahatano Village at the Izu Peninsula, large Myrica trees form a part of the upper forest canopy, but young trees are not found inside the forest. On the other hand, in the inland zone forest, where management has been practised to obtain evergreen and deciduous coppice forests, young trees are found at high frequency. This suggest that the availability of more light through cutting has made invasion and growth of Myrica trees possible (Kajiura 1990).

To be able to follow how a population develops over time, it should be monitored regularly. The following activities are part of the monitoring process: defining objectives, identifying key associated taxa and sample quadrat locations, selection of data for quadrat recording, determination of desirable frequency and timing of quadrat recording, accumulation of data sets, statistical analysis and production of recommendations on the management plan (Maxted et al. 1997).

Since it is impossible to record and monitor all species or individual plants, samples are taken and key indicator taxa, mostly including the target species, and key indicator sites are measured. This involves the use of both fixed and random quadrats or transects within the reserve. There are several methods to measure the abundance or diversity within quadrats or transects. The most commonly used is presence or absence with an estimate of density when present. Absolute numbers could also be taken, but that is more time consuming. There is no recommended size for quadrats, but it is the best to keep the same size for all quadrats at every location and time. Furthermore, monitoring should take place in the same season to facilitate comparisons from year to year.

Reserve Utilization

Genetic reserves are not just established for conservation on itself, but to conserve PGR which might go extinct and be needed in the near or distant future. The conserved material is or could be utilized by different groups and for different reasons. Three different groups of users are recognised:

1. local communities.
2. general public.
3. scientific professionals.

The establishment of a reserve will often disrupt the normal activities carried out by local people, such as collecting, hunting or simply visiting the area. To work in the reserve in a sustainable manner, the support of the local community for conservation should be obtained. To avoid disruption to the activities of local communities, wherever possible, sustainable use of the buffer and transition zones by local traditional communities should be encouraged. The second user group is the general public, who pay for the reserves by taxation. So it is very important to have the support of the general public and make them aware of the importance of conservation. Both ethical and aesthetic reasons are often very well accepted. In addition, the reserves could be used by the general public for example for education and eco-tourism.

The scientific professionals will utilize the conserved material in the same way as material that is conserved ex situ. A disadvantage of the in situ approach might be that the material is not easily accessible for the plant breeder, this can be avoided by characterizing much of the material and disseminate the collated information. The seasonality of the reserve might also limit the access of the reserve, because fruits and seeds can only be collected during the fruiting season of the species. An advantage of in situ genetic reserves for the professionals is that evolutionary studies of the target species can be carried out.

Examples of Genetic Reserves

Myrica rubra is dioecious with its female tree bearing reddish to purple fruits as well as a dye from its bark. It is distributed in coastal areas from southern China to central Japan. Fruit of M. rubra is consumed fresh or brewed into wine or reddish liquor. Ukiyamma, a forest located on a flat lava terrace on the Izu Peninsula in Japan is used as a genetic reserve for M. rubra. In the 19th century, the management of this forest was directly controlled by the Samurai government. Harvesting of the mature fruits of M. rubra in the forest was permitted only for inhabitants. If anyone cut a trunk or even a shoot, the death penalty by decapitation was imposed because M. rubra was an important economic crop at that time. Since then, M. rubra trees have been preserved as common property by means of an agreement by the villagers (Kajiura 1990).

Kostermans and Bompard (1993) recommended that conservation of Mangifera species should be both ex situ and in situ. Although in situ conservation has advantages, since the trees remain in their natural habitat, its big drawback is that the nature reserves in Asia, especially in Indonesia, are not secure. From this example, it is clear that further investigations are needed to explore if in situ conservation is possible or not.

In 1993, a pioneering project on in, situ conservation of wild crop relatives and important forest trees species was started in Turkey. This project aims to provide a national strategy and action plan for the conservation of wild relatives through the development of in situ conservation areas (called management zones) where populations of the most important wild relatives occur. It involves ecogeographic surveys, selection of the sites, development of management plans, and activities to maintain the selected populations, creation of an effective national plan and capacity to carry out the planned programme on a sustained basis (Hodgkin 1996).

At the moment, seven 'Biosphere reserves' have been established in India by the Department of Environment and Forests to preserve biodiversity. There are plans underway for the establishment of another seven. These reserves have been chosen, because they all have a high genetic diversity and also because they represent ecosystems in the different biogeographic regions of the country.

In relation to tropical fruits, it would be more important to have information on the biodiversity held within the core areas of a reserve. Until detailed surveys are carried out in the biosphere reserves, it is rather difficult to have exact information on the diversity of fruit crops. The only information available can be found in the synthesis published by the National Bureau of Plant Genetic Resources (NBPGR) in India (Arora and Paroda 1991).

In the Western Ghats, where the biosphere reserve of Nilgiri Hills is located, jackfruit and mango are known to be distributed. Furthermore, species of Syzygium are found in the Eastern Ghats, where the Gulf of Mannar biosphere reserve is located. Lastly, various Citrus species are found in the northeastern region where the Manas biosphere reserve is located and the Namdapha and Kaziranga biosphere reserves are to be established. Even though information is collected on the presence of fruit crops, no data are available on the density and diversity of each species.

On-farm Conservation

As defined earlier, on-farm conservation of PGR is concerned with crops, and their weedy and wild relatives. On-farm conservation is dynamic and is aimed at maintaining the evolutionary processes that continue to shape genetic diversity. It is based on the recognition that farmers have improved and grown genetic diversity and that this process still continues among many farmers in spite of socio-economic and technical changes.

Farmers play a big role through their selection of plant material which influences the evolutionary process and through their decisions to continue with a certain landrace or not (Bellon 1995).

Landraces are part of agricultural systems. Therefore, on-site conservation cannot be achieved through their isolation in biological reserves. According to Brush (1991) and Bellon (1995), farmers should be encouraged to continue planting their landraces and thereby conserve them. This implies a radically different approach to conservation, one in which farmers play a role equally important as that of scientists and politicians (Bellon 1995). There have been only a few studies aimed at studying on-farm conservation and up to now, no model exists for the development and establishment of on-farm conservation. If on-farm conservation is to be implemented, a fuller understanding of both crop populations on the farming systems that produce them is needed to create active cooperation between farmers and conservationists (Brush 1991, 1995). Bellon (1995) argues that the focus of on-farm conservation should not only be on the varieties a farmer grows, but also on seed flows, variety selection, variety adaptation and seed selection and storage. All elements of the farmers' management influence the whole process of on-farm conservation and these should be studied.

Since scientific back-up for on-farm conservation is lacking, Brush (1991) suggests five general guiding principles that may be used in evaluating proposals and pilot on-farm conservation programmes:

1. Even while focusing on on-farm conservation, it is important to be aware of the complementarity of the different techniques. Ex situ conservation be used as a back-up in case material is lost in situ.
2. Institutional development of in situ conservation strategies should be related to the activities already found in. farming system.
3. The existing institutions and incentives should be reinforced rather than create new ones to provide continuity.
4. Conservation by farmers can be strengthened by agricultural development policies that enhance incentives to continue to maintain landraces. Given the role of farmers in on-farm conservation, meeting development goals such as increased farm income, is critical.
5. Crop germplasm is an international public good and its conservation must be supported through international means.

According to Brush (1991), the main challenge in implementing in situ conservation for crop species is to encourage farmers to continue cultivation of the various landraces. Therefore, he suggests that an institutional framework, an extensive information base and a policy framework are required to develop and establish an on-farm reserve.

IPGRI, based on a number of Brush's recommendations, has developed a global project entitled "Strengthening the Scientific Basis of in Situ Conservation of Agricultural Biodiversity". The project consists of case studies in 9 countries including Nepal and Vietnam. The objectives of this project are to:

1. Enhance and support a framework of knowledge on farmer decision making processes that influence in situ conservation of agricultural crop diversity.
2. Strengthen national institutions in planning and implementation of conservation programmes for agricultural biodiversity.
3. Broaden the use of agricultural biodiversity and to involve farming communities in its conservation.

As a first step, a basic data set will be collected through interviews on farmer decision making in the selection and maintenance of landraces to provide measurable indices of genetic diversity of the landraces. In addition, morphological and agricultural information relating to characteristics that the farmer does not actively select for are also being gathered. Finally, biochemical and molecular markers will be used. Once a basic data set is established, key factors that affect farmer decision making will be identified. A wide range of environmental biological, socioeconomic and/or cultural factors will be included. To obtain all the information described above, IPGRI collaborates closely with national programmes. To support national capacity building, IPGRI will also provide training on in situ conservation in the fields of conservation biology, populations genetics, ecology, etc.

Broadening the use of agricultural biodiversity will be based on the following activities:

- Decentralised farmer managed selection and crop breeding.
- Market and non-market incentives, resource tenure and promotion of local plant uses.
- National policies which enhance maintenance.

Participation of farming communities and the informal sector is very important in these activities. Lastly, it is important to collect information about the sustainability of on-farm conservation. This could be done on-basis of following questions (Brush 1995).

- Will farmers continue to maintain landraces even if land distribution systems change?
- In many places, landraces are still conserved, however, in a much smaller scale (Brush 1991, 1995). How will this reduction affect crop evolution?
- What influence has human population growth on on-farm conservation sites?

All of the limited and fragmented information available on on-farm conservation is related to main food crops, such as cereals and legumes. Very limited information exists on the on-farm conservation of tropical fruits. However, two on-going IPGRI projects in which tropical fruits play a role are described in the following examples.

Examples of on-farm Conservation

Part of an on-going project on "Diversity Conservation and Sustainable Use of Genetic Resources of Fruit Trees Native to Tropical America" deals with on-farm conservation of Sapotaceae handled by IPGRI Americas office. Sapotaceae, large trees producing edible fruits, are cultivated throughout Central and South America in home gardens producing fruit for domestic consumption and additional income. Since Sapotaceae have recalcitrant seeds, in situ conservation will be an important component of any conservation strategy.

The following activities are undertaken to promote on-farm conservation. Firstly, monitoring of genetic diversity and genetic erosion are studied. Then information is gathered on farmers' decision-making. To promote the use and on-farm conservation of Sapotaceae, agronomic and post-harvest techniques are being developed, advanced materials selected and socio-economic studies on production including marketing conducted.

A pilot study on home gardens is being carried out in Vietnam to evaluate the contribution of home gardens to in situ conservation (Hodel 1997). Since several tropical fruits are grown in home gardens, this on-going study might offer useful information that could be useful to other projects on on-farm conservation of tropical fruits.

Conclusions

There is a high recognition of genetic reserves and on-farm conservation at present. Therefore, these techniques should be developed further, especially with regard to measuring genetic diversity over time, integrating the various sciences relevant to in situ conservation, developing generic models, and developing local, national and international policies. In situ conservation appears to be promising for tropical fruit crops. Results from different on-going IPGRI projects, especially those relating to tropical fruits, will give better insight as to how to develop and establish future in situ projects.

Complementary Conservation

Crop genetic resources collections are assemblies of genotypes or populations representatives of cultivars, genetic stocks, and related wild and weedy species, which are maintained in the form of plants, seeds, tissue cultures, etc. (Frankel and Soule 1981). Functionally, such plant genetic resources (PGR) include landraces, advanced/improved cultivars, and wild and weedy relatives of crop plants. Landraces may contain co-adapted gene complexes that have evolved over decades (Harlan 1992) and are the most important of the PGR. Advanced cultivars, genetic stocks and wild relatives of crop plants have a part to play in future crop improvement and, therefore, need to be preserved (Frankel 1990).

The great wealth of genetic diversity remaining in the genepools retains vast potential for present and future use. Broadly speaking, PGR are non-renewable and it is essential that we should be concerned with conservation, be it at species level, genepool level or at the ecosystem level. The limitations of a narrow genetic base of many present day varieties have been stressed many times. Genetic diversity is a defence against the genetic vulnerability, which has been built into the genetic structure of traditional cultivars. Countries which still have a significant amount of genetic diversity and species diversity have a responsibility to themselves as well as to the world at large to conserve it and make it available for use (Ramanatha Rao 1994).

Conservation Approaches and Methods

It is well known that there are two approaches to conservation of PGR - ex situ and in situ. Ex situ conservation approach generally comprises the following methods:

seed storage, field genebanks, in vitro storage, pollen storage, DNA storage and botanical gardens. Conservation of plant diversity using reserves/protected areas, on-farm and home gardens is considered as in situ conservation approach. We will now look at each method briefly.

Ex Situ Conservation

Conservation of Seeds : In the past, many collections were maintained without the help of storage which would extend the viability of seeds. Due to this, the conserved accessions had to be regenerated very frequently leading to loss of genetic diversity in genebanks (Frankel and Hawkes 1975). Generally speaking, seed storage conservation method is an efficient and reproducible technique. The seeds are dried to a lower moisture content and stored at a low temperature. This method is almost universally applied to the orthodox seed species.

However, the procedures followed in conservation, maintenance and storage present many problems. In maintaining genetic purity of the conserved accessions, problems arise due to differential survival in storage, selection during regeneration, outcrossing with other entries and genetic drift (Allard 1970). Good storage conditions coupled with proper grow-outs are expected to reduce the effects of such problems (Rao 1980). It is well recognized that seeds should be maintained under conditions in which the life processes in seeds are minimized so that they can be stored for a number of years with little loss in genetic diversity, genetic integrity and viability. In general, limits can be set for loss of viability as indicator of the genetic shifts in the stored seed material.

Due to the extended life span of seeds stored under optimum conditions (cool and dry), it will not be necessary to regenerate the seeds at frequent intervals. Seeds of most common species can be maintained in this way for a number of years. Reduced frequency of regeneration results in more cost-effective maintenance of germplasm. More importantly, it minimizes the genetic erosion resulting from genetic drift in small populations that may be grown for regeneration; genetic shifts resulting from natural selection when material is grown out in the field (especially when grown under completely different conditions from that of the original location) and the risk of human error (Frankel and Hawkes 1975; Ramanatha Rao 1991). Guidelines for proper handling and storage of seeds of many different crop species are now available.

Work is also in progress on alternative methods of storage of seeds such as the maintenance of seeds imbibed storage, storage of seeds in liquid nitrogen at temperatures below-196°C and storage of ultradry seeds. As opposed to common seeds, generally called 'orthodox' seeds which can be dried to very low levels of seed moisture content (below 7%), there are a number of species whose seeds cannot be dried to low levels for optimum storage. Such seeds have been referred to as 'recalcitrant' (Roberts and King 1986) and imbibed storage (at higher levels of seed moisture) may be of considerable importance. Efforts are also underway to undarstand the genetic basis of recalcitrance which will be necessary to deal with this problem. Very low temperature storage using liquid nitrogen, called cryopreservation, also appears to be promising with a more extended life span than seeds stored in currently what is described as long term storage (-20°C). Another area in which considerable work is required is on storage of ultradry seeds (dried to seed moisture content of 2-5%) at room temperature conditions. It is now known that such low levels of seed moisture content would help greatly in extending the seed longevity under ambient conditions, when stored in hermetically sealed containers. However, more research will be necessary before ultradry seed technology can be adopted.

Conservation of Plants

Many important varieties of field, horticultural and forestry species are either difficult or impossible to conserve as seeds (*i.e.*, no seeds are formed or if formed, they are recalcitrant) or reproduce vegetatively. Hence, they are conserved in field genebanks (FGB). FGBs may run a risk of being damaged by natural calamities, infection, neglect or abuse. Ex situ conservation of tree species using FGBs requires a substantial number of individual genotypes to be an effective conservation measure.

Thus, FGBs require more space, especially for large plants such as tree species, and they may be relatively expensive to maintain depending upon the location and the complexity of alternative techniques available. However, FBGs provide easy and ready access to conserved material for research as well as for use. For a number of plant species, the alternative methods are not fully developed yet so that they can be effectively used (Ramanatha Rao and Riley 1995; Ramanatha Rao et al. 1996). Thus, it is clear that establishment of FGBs will play a major role in any conservation strategy for PGR. It is one of the options of a complementary strategy for the conservation of germplasm of many plant species. At the same time, efforts to develop and refine other methods, such as in vitro conservation and on-farm conservation must continue.

Conservation of Tissues/cells

Another method of ex situ conservation of germplasm is through the use of tissue cultures. There are a number of important plant species which cannot be conserved as seeds and present different problems. Generally, conservation of such material as plants in the field requires greater space, labour and costs. They also run the risk of being damaged by natural calamities and virus infections as noted earlier. Hence, the conservation of tuber, root, shrub and tree species becomes very difficult indeed. Several techniques to conserve such vegetatively propagated species have recently been developed and some of them are undergoing rigorous testing. Possibilities now exist to conserve PGR as tissue cultures. For some species, the in vitro conservation may be the only option available.

Though tissue culture offers great potential for conservation of germplasm of vegetatively propagated material, two things have been of major technical hindrance to it. Firstly, the genetic instability of the material conserved as tissue culture due to somaclonal variation at the time of regeneration of the tissue into seedlings. Secondly, the length of storage as tissue has been limited.

Significant work is being done on both aspects and for some species, tissue culture maintenance is relevant due to improved techniques resulting in low levels of somaclonal variation. Work on cryopreservation of tissue culture so that these can be preserved for long periods is also making rapid progress. Once these techniques are further refined, their large-scale adoption will be possible.

Conservation of Pollen

Pollen storage was mainly developed as a tool for controlled pollination of asynchronous flowering genotypes, especially in fruit tree species (Alexander and Ganeshan 1993). In addition, pollen storage has also been considered as an emerging technology for genetic conservation. Even if it may not be considered to be a viable method for meaningful genetic conservation of genotypes, cryopreservation is likely to be more successful than other storage techniques routinely employed for pollen (*e.g.*, under organic, desiccation freeze drying, low temperature) in facilitating hybridization when flowering is asynchronous or for use in next season. Thus, it can help in better utilization of available genetic resources. Pollen can be easily collected and cryopreserved in large quantities in relatively small space. In addition, exchange of germplasm through pollen poses fewer quarantine problems compared with seed or other propagules. In recent years, cryopreservation techniques have been developed for pollen of an increasing number of species and cryobanks of pollen have been established for fruit tree species in several countries (Alexander and Ganeshan 1993).

DMA Storage

Storage of DNA is, in principle, simple to carry out and widely applicable. The storage of DNA seems to be relatively easy and cheap. The progress in genetic engineering has resulted in breaking down the species and genus barriers for transferring genes (NRC 1993). Transgenic plants have been produced with genes transferred from viruses, bacteria, fungi and even mice. Such efforts have lead to the establishment of DNA libraries, which store total genomic information of germplasm in the form of DNA libraries (Mattick et al. 1992). However, strategies and procedures have to be developed on how to use the material stored in the form of DNA. Therefore, the role and value of this method for PGR conservation is not completely clear yet.

Botanical Gardens

There are about 1500 botanic garden and arboreta in the world (WWF-IUN-BGCS 1989). The objective of most of the gardens are said to be to: (i) maintain essential ecological processes and life support systems, (ii) preserve genetic diversity, and (iii) ensure sustainable utilization of species and ecosystem. However, the botanical gardens may play a limited role in the context of conservation and propagation, and probably a greater role in public awareness and education.

Botanical gardens may mainly be used to display a great number of different and exotic species. As the numbers that can be maintained in this manner is limited, it can not reflect or conserve genetic diversity. There is a possibility that a few well managed gardens can lay emphasis on conservation of certain groups of species as living collections (*i.e.* field genebanks).

In Situ Conservation

This type of conservation is dynamic as opposed to the semi-static nature of ex situ conservation. One of the reasons given for choosing in situ conservation over ex situ is the need to maintain the evolutionary potential of species and populations (Frankel 1970; Frankel and Soule 1981; Ledig 1988, 1992). However, given the fact that human activities can cause habitat destruction and loss of biodiversity in some cases, and the maintenance of biodiversity in other cases, the need to complement it with ex situ conservation efforts is well recognized.

In general, research and monitoring is needed at three levels for successful in situ conservation: the assay of genetic variation represented within a target species in a particular area (ideally by studies of intraspecific morphological and molecular variation and the diversity as recognized by local users including farmers); regular inventory of species numbers; and observation of general ecological condition and habitat alteration including farming systems (Berg 1996).

Biosphere Eeserves/protected Areas

In general, the biodiversity at the species and ecosystem level can only be conserved through in situ conservation (McNeely 1996). Various types of protected or semi-protected areas that are identified to be rich in diversity of ecosystems and/or species are used in this method. Conservation of wild species crop relatives in genetic reserves involves the location, designation, management and monitoring of genetic diversity in a particular, natural location (Maxted et al. 1997). However, it must be remembered that genetic reserves are often not very accessible for use. Additionally, if the monitoring and management may not be optimal due to difficult conditions under which these need to be performed. For the same reason, characterization and evaluation will be limited. The reserves are also vulnerable to natural and human-made disasters.

On-farm Conservation

In situ conservation of agrobiodiversity or on-farm conservation involves the maintenance of traditional crop cultivars (landraces) or farming systems by farmers within traditional agricultural systems. Traditional farmers use landraces, which are developed by the farmer and strongly adapted to the local environment (Harlan 1992). This method of conservation has been gaining importance in recent years, though farmers have used it for centuries.

In the case of agrobiodiversity, the effects of growers-practices are of paramount importance. In almost all such areas, there is presently little information available on the status of the genetic diversity. It is now possible to monitor and estimate genetic diversity using molecular markers. However, the limited resources available for such work makes such a proposal difficult to implement. Measurement of genetic diversity in most perennial tree species depends largely on morphometric traits, but increasing use of molecular markers can assist in better understanding of the structure of genetic diversity both at a specific site and across regions.

Systematic documentation of farmers' knowledge of diversity and uses is needed. Sustainable in situ conservation will require community participation, control of land rights in local communities, education, extension and development of environmental awareness. Of equal importance is the principle that any in situ conservation programme must benefit the local communities. Management by local communities can often be developed to effectively link conservation and use (McNeely 1994, 1996). It is important to consider indigenous knowledge, peoples participation and cooperation between local people, researcher and conservationists and non-governmental organizations (NGOs). Additionally, it is important to consider the establishment of areas of intensive management or high yielding plantations for long term sustainability of any in situ conservation programme. Conservation activities by commercial and private agencies can also be promoted as these groups have the capacity to fund such activities. Since it will be necessary to foster sustained conservation and use of resources to derive long-term benefits from the exploitation of the resources, we believe that the commercial sector and private agencies will be interested in activities mentioned above. This can lead to much wanted linkages among public, community and private sectors in plant genetic resources conservation (Riley 1995). The methods of management and benefits to local communities in maintaining and using this diversity must be considered while implementing an in situ (or on-farm) conservation programme.

Home Gardens

Home garden conservation is very similar to on-farm conservation, however, the scale is much smaller. In most rural situations, home gardens tend to contain a wide spectrum of species, such as vegetables, fruits, medicinal plants and spices, than on-farm plots. As it is akin to on-farm conservation, the dynamic nature of this conservation technique has same advantages. Home garden, as a single unit, has very little value in terms of conservation, but a community of them in a given area may contribute significantly to the conservation and direct use of genetic diversity. Most of such diversity could be somewhat unique/rare as the people tend to grow unique materials in their gardens and also underutilized species or even un-domesticated species. However, the system is vulnerable to changes in management practices. Home gardens are also known to be testing grounds for farmer-home gardener as well as a location for testing out some of the wild and semi-wild species. Thus, in rural areas, the home gardens will continue to play a role in genetic diversity conservation as well as development.

Complementary Conservation Strategy

It is important to emphasize that these two main approaches to conservation of PGR: ex situ and in situ are complementary in nature. Conserving a genepool should employ a combination of methods, from nature reserves to genebanks as no single method can conserve all the diversity. The appropriate balance between different methods employed depends on factors such as the biological characteristics of the genepool, infrastructure and human resources, number of accessions in a given collection and its geographic site, and the intended use of the conserved germplasm. For any given genepool, the extent of a particular method used may defer from that used in another genepool (IPGRI 1993).

Genepool

Focusing conservation on genepool basis appears to be the best approach, since it will help us to conserve maximum genetic diversity of any group of crop or forestry plant species (IPGRI 1993). So for effective collecting and conservation, it is important to determine the genepool of species. Most commonly three types of genepools have been distinguished (Harlan and de Wet 1971):

- *Primary Genepool:* This corresponds with the botanical species. It consists of cultivated species and its weedy and wild relatives, which can cross with ease among themselves and produce fertile F1s.
- *Secondary Genepool:* This consists of phylogenetically related species, which can be crossed with the species of interest, and may produce partly fertile, but weak F1s.
- *Tertiary Genepool:* As the name indicates, it consist of species that are phylogenetically related and normally do not intermate. However, they may be crossed using complicated techniques, producing weak and often sterile F1s.

Except monotypic species, all others are expected to have relatives that fall into this classification. With the advent of modern biotechnological techniques that can transcend species/genus barriers, the definition of genepools is slightly blurred, but this concept still helps us to group the target species and its related species into understandable divisions. Once the target species and the genepools are determined, steps in developing complementary conservation strategy can be taken. To formulate a conservation strategy, the conservationist need to address a number of questions in relation to the genepool. In the beginning, we have noted that any complementary conservation strategy depends on factors, such as the biological characteristics of the genepool, infrastructure and human resources, number of accessions in a given collection and its geographic site and the intended use of the conserved germplasm. These factors can be used to formulate the following questions. These are not exhaustive, but cover most of the ground.

Biological characteristics Genetic diversity:

- What is known about the genetic diversity within the genepool, at each level?
- What special traits or unique diversity should one be looking for?

Life Cycle:

- Does the genepool include only short lived or long lived or a mixed group?
- What is the age of the plant when it becomes useful for either breeding or for other purposes?

Reproductive Biology :

- Do all species under consideration reproduce sexually or asexually, or a combination of the two?
- If reproduction is sexual, then is it a selfing or outcrossing species?
- What is the level of outcrossing?
- Are there any other biological traits that will interfere with true seed production?

Storage Characteristics:

- Are species under consideration orthodox or recalcitrant seed species?
- Should the storage be for short, medium or long-term?

Location and Facilities :

- Where is the germplasm located?
- Can all the species in different genepools be maintained or grown in the same location?

Infrastructure:

- How accessible is the area of collection?
- How accessible has the material to be and for whom?
- What facilities for different types of conservation are available?

Human Resources:

- What is the level of training of the staff?
- Can the level be upgraded periodically?

Objectives of Collection:

- Is the aim to conserve genes or genotypes?
- Will it completely stop evolutionary processes?

Improvement Strategy:

- Is the collection linked to the user community?
- What kind of breeding methods are used?

Others Factors—Legal Issues:

- What legal arrangements exist for transfer and access to genetic material?
- What are the legal issues that govern the acquisition of land and other related facilities?

Importance of the Crop:

- How important is the crop?
- Is it important enough for the local people to participate in situ conservation effort?

Model Complementary Conservation Strategy

If most of the above questions are answered, then an appropriate conservation method can be determined. A combination of methods, from both the approaches, which are complementary, should be applied. Such a combination should balance amount of germplasm that can be conserved using each method so as to provide us with a strategy that optimally conserves maximum diversity within the target genepool.

For example, if one considers the conservation of Citrus genepool, at the simplest level, one has Citrus species that have orthodox seeds or intermediate seeds or recalcitrant seeds. All other things being equal, one could say that the Citrus species that produce orthodox seed are conserved in seed genebank (However, this can be complicated by many other considerations, for example, objective of conservation, polyembryony, etc.).

Species that produce recalcitrant seeds, depending on the level and stability of the technology available may be conserved as tissue (slow growth or cryopreservation) or in field genebanks. If the decision is to use former, the next question to answer is what is the gestation period for in vitro conserved material to become useful. In this fashion, it will be possible to go through the list of questions and determine what part of the genepool will be conserved using which method, thus establishing a complementary conservation strategy for Citrus. As one can see, there is a need for complete information to make the appropriate choices. Personally, I think that attempts to develop models at species group level may be futile as the number and type of parameters considered for any one genepool may vary greatly, and these will make any model effectively unworkable.

Concluding Remarks

No single method can help us to conserve most of the genetic diversity in any given crop or genepool. What we need is a complementary conservation strategy which make, use of optimally all techniques/methods available (and leave place for any that need to be developed or refined). It is a process of making decision about the conservation of a genepool. Such a strategy should be able to meet most of the needs in PGR conservation.

It should be pragmatic and at the same time critical, especially in determining the methods to be used. It does not advocate methods, simply because the method is available, but only because it is the most appropriate one under the given conditions. A good complementary conservation strategy does not categorize crops or species into definitive classes. It is dynamic, and lends itself to meet the challenges of changes that are occurring in the field of genetic resources as it is open to new technologies and new needs.

4

Management of Fungal Diseases in Tropical Fruits

In India, the diverse agro-climatic conditions favour cultivation of more than five kinds of fruits and nut crops, majority of which are grown in the tropical region. The major commercial tropical fruits are grapes, banana, mango, citrus, papaya, pineapple and sapota. Although the area under fruit crops has increased in the last three decades, but still it is only a little over one per cent of the total cultivated area.

The increase in area alone cannot maximise the yields unless and until per unit area production is increased. It is possible only through the development of high yielding varieties and the knowledge about the factors responsible for the crop losses. Among these factors, plant diseases specially fungal diseases, play a major role in reducing yields of horticultural crops in the tropics.

It is estimated that the production could be increased at least by 28% if the crop could be protected against various diseases. Many a time, most of the diseases lead to the complete crop failure if it is not checked in time. Many of these diseases are transmitted through the planting material.

The phenology of fruit crops which are perennial is different from other field crops which are seasonal in nature and hence the pathogens keep on perpetuating on the plant itself in fruit crops.

Further, the knowledge about etiology and epidemiology of diseases is most important which would help in profitable cultivation of fruit crops and the measures adopted thus, in turn, will improve the economy. This knowledge is also important for optimal management of field genebanks of various fruit species. The available relevant literature on some of the most important crops has been screened and is presented here.

Fungal Diseases in Tropical Fruits

Mango

Mango (Mangifera indica L.) is an important fruit and is subjected to a number of diseases at all stages of its development *i.e.* from nursery to the consumption of fruits. All the plant parts, namely, trunk, branch, twig, leaf, petiole, flower and fruit are attacked by different pathogens. Some of the diseases are very severe on plant and produce alike and have become a limiting factor in profitable mango orcharding. Bloom blight caused a complete failure of the crop in Florida and Brazil, black spot in South Africa and powdery mildew and Black-tip in India, etc.

Many types of agents cause diseases in mango and of these fungi cause the largest number of diseases, while bacteria, algae, angiospermic parasites and nutritional deficiencies are the other causal agents of mango maladies. Virus, which causes severe infection in other fruit crops, is luckily not reported so far on mango.

Many of the mango diseases have been investigated and also measures to combat them have been worked out in different countries. However, our knowledge about them is still imperfect and inadequate. Vast areas in South-East Asia - the home of mango, still remain unexplored from disease point of view and may possibly offer many other problems for investigations. The information available on some of the important diseases is presented below.

Powdery Mildew (Oidium mangifera Bert.)

First reported from Brazil in 1914 by Berthet, later on it has been recorded from several countries of the world. In India, it has been reported from Uttar Pradesh, Punjab, Maharashtra, Gujarat, Madhya Pradesh, Jammu & Kashmir, Haryana, Andhra Pradesh, Karnataka and Tamil Nadu. The losses caused vary from 5-20 per cent depending upon the weather conditions. Prakash and Srivastava (1987) reported 30-90% losses in Lucknow. This is a serious problem in hills, valleys and tarai regions of Uttar Pradesh. Serious outbreaks have been noticed in Uttar Pradesh in the last 6-7 years. It attacks the leaves, flower scales, buds of tender flower heads, axils, stalks and fruits. It manifests itself as wefts of white mycelium on the affected parts.

The whole surface is later on covered with a powdery substance which is blown away by even a slight disturbance caused by winds. The fungus is ectophytic, remaining mostly on the surface and draws its nutrition from the plant cells through the haustoria which penetrate the epidermal layer and reach the cells. The affected fruits do not grow in size and may drop before attaining pea size. In Bangalore area, the disease has been found to be prevalent almost throughout the year on varieties like Pairie and Alphonso.

In Indo-gangetic region the infection is occasionally found on malformed inflorescences and there is a likelihood of its survival in the sub-mountainous regions of Uttar Pradesh. Warm temperature with heavy morning dew and cloudy weather favour the disease development (Kulkarni 1924). The minimum, optimum and maximum temperature for conidia germination is stated to be 9, 22 and 30.5°C respectively (Uppal et al. 1941). The disease appearance and incidence is greatly influenced by the sunshine (hours) per day and it is inversely proportional to the disease having a correlation coefficient of r =0.5943 (Gupta 1985). The other factors influencing the disease was relative humidity at 1430 hours (Rawal and Saxena 1992).

The most effective method of control is by dusting the plants with fine sulphur (250-300 mesh) at the rate of 1-3 lb. of sulphur/tree (Cheema et al. 1954). The first application may be soon after flowering, second 15 days later followed by a third one. Spraying of Guesarol (405.50) has been found successful in Tamil Nadu, Andhra pradeeh and Maharashtra. Several fungicides, such as Cosan (0.2%), benlate (0.2%), wettable sulphur and DDT (2:1), karathane or morestan, carbendizim (0.1%), 0.2% microsul and butrimate, Baycor, Calixin, Anvil (0.05%), Systhane (0.05%), Saprol (0.2%) were found to be effective in controlling the disease.

Anthracnose: This is caused by Colletotrichum gloeosporioides Penz. = Glomerella cingulata (Sten.) Spould and Shrenk., Gloeosporium mangiferae P. Henn and is the most important disease wherever mango is grown. The disease develops on all the tender parts of the plant and is especially serious on tender twigs, flowers, flower stems and fruits in storage.

In India, it is widely distributed in Bihar, Punjab, Maharashtra, Gujarat, Kerala, Uttar Pradesh, Rajasthan and Karnataka. The disease produces leaf spot, blossom blight, wither tip, twig blight and fruit rot symptoms. Young leaves when attacked wither and dry up. Sometimes only the edges of the leaves are attacked; their margins darken, dry up, may fall out, giving the leaf a rugged appearance. It produces a wither tip of young tender twigs. It also results in a die-back which appears as blackish spot on the growing tips. The affected branches ultimately dry up and infections keep on penetrating down.

On fruits, most of the infection takes place from the start of the blossoming period until the fruits are more than half grown. The spots appear near the stem end as small brown areas that enlarge rapidly and become black. In some cases, the areas involved are in the form of streaks running down from the stem end. The affected areas usually crack and sink slightly.

The decay is confined to the skin of the fruit except in late stages when it penetrates the flesh in shallow areas. The latent infection (during ripening) is carried from the field and develops further which causes rotting in the storage. Healthy fruits develop infection after coming in contact with diseased fruits. The inoculum remains on dried leaves, defoliated branches and mummified flowers and flower brackets. The severity and prevalence of the disease is influenced by excessive rains, heavy dews or high humidity during critical infection period (Doidge 1932). The temperature range for disease development 10-30°C, and the R.H. 95-97% are highly congenial. Spotting of leaves and twigs which become severe in damp weather are generally not serious enough to cause wither tip and die back in full grown trees (Kausar et al. 1960). The secondary spread is through rain drops.

Only one variety, namely, Edward has been reported to be resistant (Brooke and Olmo 1957). In Uttar Pradesh, anthracnose was effectively controlled by spraying with carbendazim (1%) or Topsin-M (0.1%) or chlorothalonil (0.2%) at 14 days intervals until harvest. Benlate (0.2%) and Dithane Z-78 (0.2%) are extremely toxic to fungus in culture.

However, these have not been tested in the field (Bose et al. 1953). Spraying of Micop, Blitox and dithiocarbomate were found effective (Lingaraj 1969). Bordeaux mixture (1.0%) has been recommended (Gadre 1979). Application of Captan (3 g/l) or Phaltan (2g/l) are also recommended (Oriux 1961).

For the control in storage, the management strategies recommended to control anthracnose include cultural practices, tree management, varietal selection and protective sprays using curative fungicides with advantage that sprays need only to be applied after infection occur and using prochloraz and disease detection instrument known as mango anthracnose estimator (MAE) (Fitzell and Peak 1985). Hot water treatment has also been recommended by various workers which is 51.5°C for 15 minutes (Pennock and Maldonaldo 1961); 50.5-55.5 °C for 5 minutes (Soot and Segall 1963); 54-55°C for 15 minutes (Tandon and Singh 1968) and 55°C for 5 minutes added with Benomyl or TBZ (Spalding and Reeder 1972).

Anthracnose would be controlled by dipping the fruits in various fungicides or exposing them to ammonia, sulphur dioxide and carbon dioxide gases and also by hot air treatment (Tandon and Singh 1968). The fruit dipping in benomyl solution (9500 ppm a.i) or thiobendazole (1000 ppm) immediately before storage reduces the disease to 5 per cent from 29 per cent (Sohi et al. 1973). Pre-harvest spraying with Bavistin (0.1%) or Topsin-M(0.1%) or Prochloraz (0.1%) followed by dipping in the same after harvesting have also been found to be effective.

Mango Malformation (Fusarium moniliforme var. subglutinans Sheldon)

It is the most important malady of mango and was first reported by Burn (1910). It causes the losses from 50-80 per cent. It is also reported to be due to other disorders like physiological and acarological. Indications of its virus nature has also been postulated (Vasudeva 1959). The maximum incidence of the disease has been recorded in the north, northwest and north-east parts of the country.

The incidence is, however, less in western and southern India. There are two types of symptoms, namely, floral malformation and witches broom or bunchy top or vegetative malformation with a proliferation of infected tissue. The flowering panicles turn into a compact mass of flowers. This compact mass is very hard and not soft like normal panicle. Individual flower is greatly enlarged and has a large disc. The percentage of bisexual flowers in malformed panicles is very low. In bunchy top, compact leaves are formed in a bunch at the apex of shoot or in the leaf axil. A similar bunch consisting of small rudiments crowded together on short shootlets is seen in vegetative malformation in which the growth of shootlet is arrested. Vegetative malformation is more pronounced in young seedling and seedling trees. The malformed heads dry up in black masses and persist on the trees for a long time. The malformed inflorescences contain more of endogenous cytokinins than healthy ones.

The variety Zebda has been reported to be resistant in Egypt (El Ghandour et al. 1979). Considerable incidence reduction has been reported by spraying 100-200 ppm NAA during October (Majumdar 1977). Use of disease free planting material and prophylactic spray of insecticides and fungicides can keep the orchards healthy. In areas with less than 5-10 per cent infection, pruning of diseased plants should be made compulsory; pruning of diseased parts along the basal 15-20 cm apparently healthy portions. This is followed by the spraying of Bavistin (0.1%) or Captaf (0.2%). Partial control of malformation has been achieved by spraying mangiferin-Zn_2 and mangiferin-Cu2+chelates.

Stem End Rot (Botryodiplodia theobromae Pat)

It is an important disease of ripe mango. This fungus also causes die-back symptoms in orchards. It is known to occur in India, Myanmar, Sri Lanka, the Philippines, Mauritius and the U.S.A. In India, it is recorded from all the mango growing areas. The disease is characterized as the dark epicarp around the base of the pedicel in the initial stage.

In the next few hours, the affected area enlarges to form a circular, black patch which in humid atmosphere extends rapidly and turns the whole fruit completely black within two or three days. The pulp of the diseased fruits become brown and somewhat softer. The diseased fruits showed increased phosphorus and reduced ascorbic acid contents (Desai and Pathak 1970). Pectolytic enzyme activity was also lowered in such fruits (Pathak and Srivastava 1969). The disease is carried from the dead twigs and bark of the trees harbouring the pathogen and with the onset of rains the orchards get contaminated.

Dipping of mangoes in 6 per cent borax solution at 43°C for 3 minutes reduces the incidence (Thomas and Dalai 1968). Fruit dipping in Aureofungin and wrapping was also found effective (Dharamvir et al. 1967). Pathak (1980) suggested to harvest mangoes on clear dry day. Hot water treatment at 53°C for 10 minutes (Quimio and Quimio 1974) and at 51 to 55°C for 15 minutes (Liu 1985) have been recommended elsewhere. Care should be taken to prevent snapping-off of the pedicels. Injury should be avoided to fruits at all stages of handling. Spalding (1982) in his experiments found thiobendazole and thiophenate methyl to be the most effective ones. Recently, Rawal and Ullasa (1988) have reported that Bavistin (0.1%) or Topsin-M (0.1%) or chlorothalonil (0.2%) spraying in the field before harvesting give very effective control.

Die-back (Botryodiplodia thiobromae Pat)

It is a descructive disease of mango and is known to occur in India and other mango growing countries. The disease is manifested by discolouration and darkening of the bark some distance from the tip of the twigs. The dark area advances and young green leaves start withering first at the base and then extending outwards along the vein. The affected leaves turn brown and the margins roll upward. At this stage, the twig or branch dies, leaves shrivel and fall, and this may be accompanied by exudation of gum.

The infected twigs show internal discolouration. Brown streaks of vascular tissues is seen on splitting the twigs lengthwise along the long axis. In early stages, epidermal and sub-epidermal cells of twigs appear slightly shrivelled. The areas of cambium and phloem show brown discolouration and yellow gum like exudate flows out of the cells.

The disease is known to be enhanced by a beetle, Xyleborus affinis (Batista 1947). Relative humidity above 80% and temperature of 25 to 31.5°C, and rains enhance disease development. The die-back is also caused due to Colletotrichum gloeosporioides. Both these pathogens also attack fruits which get spoiled during ripening.

For effective control of the disease, pruning and destruction of infected twigs is the foremost practice. Spraying the trees periodically with copper oxychloride sulphate is recommended by Alvarez and Lopez (1971). Pasting of trees with a mixture of oil and 5% phenol was found effective (Batista 1947). Sprays of carbendazim (0.1%) or methyl thiophenate (0.1%) or chlorothalonil (0.2%) at fortnightly interval during rainy season is important. Sometimes short hole borers also predispose the trees to infection and hence proper insecticides are also to be sprayed. Healthy twigs should be selected for grafting of seedling during propagation.

Leaf Blight (Macrophomina mangiferae Hingorani & Sharma)

It was first reported from Brazil to be caused by Macrophomina sp. by Lacca (1922). In India, it was reported by Patel et al. (1949). Later, it has been reported by various workers from Delhi, U.P. and Gujarat. The other pathogens causing the disease are grey blight (Pestalotiopsis mangiferae), phoma blight (Phoma glomerate), and twig blight and fruit rot (Phoma sp.).

The disease usually appears as yellowish pin head like spots on leaves and twigs of the affected plants. These gradually enlarge discolouring the surrounding tissues which first become brown and then dark brown with slightly raised and brown purplish margins and later ash coloured due to the appearance of pycnidia.

Spots are round to start with but later become oval or irregular in size depending upon environmental conditions such as humidity and temperature. Infection is mainly observed on leaves and rarely on stems. On stem, the lesions are elliptic which later girdle the stem at the point of infection. On fruits, water soaked circular lesion are produced which enlarge rapidly and cause rotting.

The pathogen can survive for more than a year on the leaves of mango (Hingorani et al. 1960). The varieties, namely, Khandeshi, Borasio and Asal Damido are reported to be resistant (Desai and Patel 1963). Field sanitation is by collecting and destruction of diseased parts. Spraying of Burgandy mixture, Perenox, lime sulphur and dithane have been recommended by Hingorani et al. (1960).

Red Rust (Cephaleuros mycoides Karst)

This or red spot is a common algal disease on the mango in the tarai and in the other humid regions of India. Its occurrence has been reported from Bihar, Karnataka and Uttar Pradesh. The alga attacks the foliage, bark and twigs of the host plant and is both parasitic and epiphytic in nature. In serious infections, the bark becomes thickened, the twigs get enlarged and remain stunted and the foliage becomes sparse and finally dries up. Initially the spots are greenish-grey in colour and velvety in texture, but later on the surface bears reddish-brown appearance. Algal spot is circular to irregular in shape, slightly elevated and with usually 2 mm in diameter, though in some cases it may be as much as 1 cm. Bordeaux mixture (6:6:100), Supramar, Fytolan, Blitox-50 or lime sulphur are recommended (Pathak 1980).

Black Spot (Xanthomonas compestris pv mangiferae indicae)

It is one of the serious diseases and is reported from many countries. In India, it is reported from Maharashtra, Delhi, Uttar Pradesh, Tamil Nadu and Karnataka. It affects leaves, petioles, fruits and tender stems. Numerous small angular water soaked lesions appear in groups towards the tip of the leaf blade.

These are first light yellowish but later turn dark brown to black and get surrounded by a distinct halo. Several lesions are often found to coalesce forming large necrotic spots. On young fruits, water-soaked lesions develop which also turn dark brown to black. Infected fruits may show skin cracking and badly affected ones dry prematurely. Infection by wounding produces diseased spots on tender stems, petioles and stalk bearing fruits. These spots turn into deep longitudinal scars accompanied by appreciable amount of bacterial gum-like ooze. The incubation period on leaves varies from 2-14 days (Patel et al. 1949).

In India, the disease remains dormant during November to March due to low temperature (11.8-22°C) and the leaf infection is considerably reduced by the fall of the infected leaves. Kent mangoes showed close relationship of rainy season with incidence of bacterial leaf spot. One day's rainfall did not affect the disease appearance, whereas four days of rainfall led to the 36.9% disease incidence.

Twig canker persists and initiates fruit infection. The disease spread is rapid during the rains and becomes severe in July-August. Five application of Bordeaux mixture (4:4:50) with spreader was recommended by Wager (1937), whereas four sprays of 6:6:50 Bordeaux mixture was found to be effective by Marloth (1947). Sprays with fungicides such as copper containing materials and agrimycin (Viljoen and Kotze 1972) have also been advocated. Shekhawat and Patel (1975) recommended orchard sanitation by way of removal of infected materials and seedling treatment as preventive measures. Agrimycin-100 proved best in arresting the disease development. Plantomycin (200 ppm) followed by agrimycin + bavistin (1000 ppm) was reported to be the best against this disease. Streptomycin sulphte followed by aureofungin have been recommended by Prakash and Raooff (1985) to control bacterial disease of mango.

Black Mould Rot (Aspergillus niger Van Teigh)

The affected fruits show yellowing of base and development of irregular, hazy, greyish spots and coalesce into dark brown or black lesions. The mesocarp of the rotted areas becomes depressed and soft. The diseased fruits show rapid decrease in ascorbic acid content (Ghose et al. 1966). The pathogen infects injured fruits (Srivastava 1968). Totapuri variety is highly susceptible (Patel 1972). A fruit dip treatment with benlate at 1500 ppm can control the rot (Bhargava and Singh 1975).

Soft Rot (Rhizopus arrhizus Fisc.)

This pathogen alone is reported to be responsible for the decay of 6.3 per cent fruits in Delhi markets (Thakur and Chenulu 1966). The pathogen incites a typical soft rot. Development of disease is slow and less intensive at low temperature (7-8°C). The disease develops rapidly between 20 and 40°C (Thakur 1972). Thakur and Chenulu (1970) found that 2-aminothiozole with 2-aminopyridine each at 7 per cent concentration protected mango fruits for 20 days.

Some of the other post-harvest diseases and pathogens are Alternaria rot (Alternaria tenuissima (Neer. Ex. fr.) Wiltshire), Sclerotium rot (Sclerotium rolfsii Sacc.), dry rot (Boothiella tetraspora Lodhi and Mirza), brown rot (Pestalotiopsis mangiferae P. Henn. and P. glandiola (Cast) Stey), fruit rot (Phomopsis mangiferae Ahmed) and black spot, Actinodichium jenkinsii (Uppal et al. 1941).

Black Tip : This disease is peculiar to India and does not occur in any other country. It has been reported to occur in regions of West Bengal, Uttar Pradesh, Bihar and Punjab in orchards located in close proximity of a brick kiln. South India is free from this disease. The symptoms manifest as small etiolated area at the distal end of the fruit, which gradually spreads, turns nearly black and covers the tip completely. Before the etiolation is complete, isolated greyish spots appear which become dark brown, enlarge and coalesce into a continuous necrotic area. The brown discolouration spreads to the neighbouring parenchyma and the deposits also appear in the ducts. The browning and deposits gradually spread throughout the mesocarp and the affected cells distintegrate and coalesce into a dead tissue. In severe cases, the necrosis extends to the endocarp.

The practical control measure lies in keeping the brick kilns away from the mango orchards. The use of telescoping chimney 12-15 m high has also been recommended (Sen 1943). Spraying of borax (0.6%) at 10-14 day intervals has been recommended (Reddy and Kapoor 1965).

Soft Nose : The problem was first observed as physiological breakdown in cvs. Kent and Haden by orchardists in 1950 in Florida. The soft nose is characterized by hallowing of the green skin in the areas between the apex and the stigmal point. A lack of firmness in diseased fruits can also be felt with experience. Upon cutting the flesh on the ventral side towards the apex of some soft nose fruits appear to be over ripe, while on shoulder and dorsal side it is unripe (Young 1957). The over ripe flesh surrounds a mass of yellowish to brown tissue which is firmer than the surrounding affected tissue and is bitter in taste.

Young (1957) concluded that the disorder is not due to the microorganisms. However, Young et al. (1963) reported the loss of only few percent in fruits harvested from trees grown on calcareous rock soil, but on acidic soil, sandy soil it is common for 15-20 per cent incidence. According to Young et al. (1963), soft nose disorder increased with N fertilization but was reduced to greater extent by increasing Ca level of the soil. Nitrogen fertilizer in sandy acidic soil increased the incidence whereas in calcareous soil there was no effect.

To overcome this disorder, the high N level of soil should not be allowed in the sandy or acidic soil and proper Ca level should also be maintained in soil. Early picking is also suggested (Young and Miner 1961).

The other disorders are little leaf caused by zinc deficiency, tip burn due to moisture stress and salt accumulation in soil and leaf scorch due to accumulation of chloride ion in leaf.

Phanerogamic Parasites (Dendrophthoe (Loranthus) viscum) : The mango trees are severely affected by this parasite particularly in neglected orchards. They are commonly present on the trunk or branches of the trees. The foliage of the infected tree is sparse, reduced in size and its bearing capacity and quality of fruits is considerably reduced. The fruits of this parasite are baccate with viscous mesocarp and attract the birds by its attractive colour. The seeds of this parasite are spread through bird and animal droppings. Such seeds, germinate and produce haustoria.

Sometime tumour like swellings are observed on the affected tissue. There are many other species of parasites which also attack mango trees. All epiphytic plants can be removed easily in early stages before their aerial roots penetrate the ground and surround the host trunk.

Some more diseases reported by various workers attacking the mango crop in the field are leaf spots due to Rhizoctonia bataticola (Taub) Butt., Cercospora mangiferae indica Munjal Lal & Chona, Alternaria alternata Fr. & Keissler, Phoma sorghicola (Sacc.) Boerema. Dorem and Venkast; Scab (Elsinoe mangiferae Bit & Jen; black band (Rhinocladium corticolum Massee); pink disease (Pellicularia solmonicolor (Berk & Br.) Dastur); root rot and damping-off (Rhizoctonia solani Kuhn) and wilt due to Verticillium albo-atrum Reinke and Berth and Fusarium solani (Mart.) Sacc. There are several more fungi causing different diseases but the above are the most important among them.

Conclusions

Mango is attacked by various types of maladies (Rawal 1990). The epidemiology of different diseases is not fully understood and hence it should be taken up on priority basis to correlate various factors playing a direct or indirect role in disease epidemics so that the model for disease prediction can be worked which will be helpful in disease forecasting. Etiology of many problems is not yet known, thus its understanding is of utmost importance. Losses caused by individual problem or a combination of problems is not reported anywhere and it has to be determined. Some of the control measure given in the literature are very old and in the process some chemicals are either not available or they are outdated and since new fungicides/ chemicals are being developed, their testing should be a continuous process to find out the best chemical for the control. This is specially important in maintaining clean and healthy field genebanks.

Citrus (Citrus spp.)

Citrus ranks the third place among the fruits grown in India and thus assumes considerable importance. It grows much better in tropical and subtropical conditions and there is an increase in area under citrus cultivation from 90.7 thousand hectares in 1961 to 279.4 thousand hectares in 1988. However, it is very sensitive to climatic changes and the varieties under cultivation vary from region to region. Sweet oranges and mandarins constitute the bulk of area under citrus fruits. Out of these two, mandarins occupy the most prominent place.

Due to their large scale cultivation, several serious problems have also cropped up. Among them the disease problems caused by many fungal, bacterial, viral, viroids and mycoplasma like organisms are important as they cause a huge economic losses and decline of the trees. Some of the fungal diseases like foot rot, root rot, gummosis, leaf fall and fruit rot, twig blight, powdery mildew and scab, etc., are the alarming diseases of citrus. In view of this, the present review is being compiled to gather more information available in the literature. Die-back or decline is the most important problem at present being faced by the country (Aiyappa and Srivastava 1967). Decline is a symptom and not a disease. Earliest symptoms are dulling of the foliage, wilting of the leaves or delayed flushing of the tree. These symptoms are usually followed by chlorosis of leaves resembling zinc deficiency, but with a difference of showing speckling with green dots of the size of pinheads to nailheads.

Trees with pinhead speckles decline faster than the trees without them. Successive flushes produced on affected branches are erect, leathery, strap-shaped and chlorotic. Small percentage of mature fruits are reduced to the size of golf balls and contain columellas and aborted seeds. As defoliation progresses twigs begin to die-back. Trees do not die completely but sooner or later are removed because they become unproductive. Many factors, such as fungi, viruses, bacteria, nematodes, insect pests, cultural malpractices, unsuitable soil pH and high water table, etc., are responsible singly or in combination for the decline of citrus in India.

Among the various agencies that cause decline (Chadha et al. 1970), the diseases caused by fungi, viruses and bacteria are more predominant. Though the malady is in existence since 18th century, it has assumed serious proportions in the last few decades. Among all the Citrus species, varieties in commercial cultivation are the most susceptible. In Florida (U.S.A.), the disease is preceded by wilt and is caused by Fusarium oxysporum Schlecht. (Timmer et al. 1979). Apart from this, the following fungal diseases are important in different parts of the country.

Gummosis, Leaf fall, Root rot, Foot rot or Fruit rot (Phytophthora spp.)

This occurs in all the citrus growing areas especially in the high rainfall areas. Even then no comprehensive survey of the disease has been made so far. There are some reports on the prevalence of this disease, viz., in South India by Rajan and Aiyappa (1944), Ramakrishnan (1954) in Maharashtra, Gujarat; and Mysore by Cheema and Bhat (1928), Frazer and Singh (1966), Kapoor and Bakshi (1967) in Punjab; and in Assam by Chowdhuri (1951).

It is caused by more than one species of Phytophthora, *i.e.,* P. nicotianae var. parasitica (Dastur) Comb, is widespread in Assam and is often associated with Fusarium lateritium Nees (Chowdhuri 1951). P. nicotianae var. parasitica is also predominant in Karnataka, whereas P. palmivora (Butl) is prevalent throughout. Phytophthora produces symptoms of decline through: (1) a rotting of the rootlets, (2) a girdling of the trunk and a dropping of blighted leaves. The plants usually blossom heavily and die before fruits mature. The fruits lying on the ground are liable to invasions by the pathogens and develop brown rot. P. citrophthora has also been reported to cause the foot and root rot.

Symptoms : Certain parts of trees are more susceptible to Phytophthora than others and the different expressions of the disease syndrome vary accordingly. The disease manifests itself in the form of large water soaked patches on the basal portion of the stem near the ground level or it may spread further. Bark on the lower part of trunk is more susceptible to infection than the bark below the ground level.

In some very wet areas, the infection is observed quite high on the trunk and even on limbs. The gumming is sometime noticed on higher branches also. Initially, the bark is firm but with ageing of the lesion it shrinks and cracks, leading to shredding in lengthwise strips as it dries. The necrotic area is at first flush with the healthy bark, but with time it shrinks and becomes sharply delimited. The live bark at the advancing margins frequently develops a callus roll that checks further spread particularly in an upward direction. Symptoms on leaves are similar to those produced by Phytophthora infestans on the foliage of potato. Blighted leaves drop, and if defoliation is heavy the impression is given that the tree is in decline. Penetration of trunk tissue may be by way of wounds or by way of growth on cracks that develop with the emergence of adventitious shoots near the bud union.

The various names given to the different types of damages caused by Phytophthora are as follows; when the infection is continued to the feeder roots it is called feeder or fibrous root rot. If the damage to the root system is more extensive and yet does not extend upwards to the crown roots it is called root rot. The damage caused to the bark of crown roots and or bark of the trunk is called Phytophthora gummosis collar rot or foot rot.

Epidemiology : Leaves and fruits may also be attacked by Phytophthora nicotianae var. parasitica during prolonged periods of rainy weather and high humidity. The attack is more serious on mandarin. Oat meal-agar with yeast or beef extract gave maximum vegetative growth, whereas for sporangial production rain water was the best (Rao 1984). Collar rot becomes more severe in black soil than in light soils (Rao 1954). P. palmivora thrives best at 25-28°C. A soil pH of 5.4-7.5 favours the disease (Aiyappa and Srivastava 1967). Mosambi and Pummelo are more susceptible than mandarin. Lime and Jambheri are resistant. According to Raychaudhuri and Lele (1970), Troyer Citrange and trifoliate oranges seem to offer protection against brown rot and gummosis incited by Phytophthora species. Invasion of host tissue is by motile, free swimming spores called zoospores. Water (not just humidity) is required for sporangial formation, dispersal of zoospores as well as for spore germination (Whiteside 1973). Brown rot was greatly influenced by the rainfall and not by spacing (Timmer and Fucik 1975).

Role of Root Stock : The predominance of resistant and moderately resistant clones was observed in pummelo, sweet lime, trifoliate orange and trifoliate orange hybrids (Grimm and Hutchinson 1977). In their earlier studies conducted in 1973, it is reported that Citrus macrophylla, C. miaray, Severinia buxifolia, Cuban shaddock, Ogami Pummelo and Sour orange were rated resistant. In Turkey, Carrizo Citrange has been reported to be highly resistant (Cinar and Tuzcu 1976).

Control Measures : Adoption of tolerant rootstocks like sour orange, Cleopatra mandarin, trifoliate orange, Rangpur lime, etc., had been advocated by Chadha et al. (1970). A variety resistant in one region may be susceptible in other. The use of nursery stock grown under Phytophthora free conditions and prevention of water standing in contact to the susceptible portions of the bark above the bud union by way of good drainage. The disease can further be prevented to some extent by dusting of walls and bottom of planting pits with zinc-copper lime in ratio of 5:1:4 (Ramakrishnan 1954).

The exposed wood can also be covered with zinc copper lime (3:1:3.5) wash or bordeaux paste which should be followed by application of white lead or any wound dressing agents (Ramakrishnan 1954). Desai et al (1966) achieved complete control of spraying Khagzi lime trees infected with P. citrophthora with aureofungin 3 g/ 30 ml of liquid soap/30 gallons of water applied 2-3 times. Somani and Patel (1970) got good control of gummosis incited by P. nicotianae var. nicotianae by aureofungin as soil drench and spraying of foliage twice at an interval of one month. Drenching of Ridomil MZ (0.2%) has also been found effective against gummosis. Captafol (3 g/l), copper ammonium carbonate and cupric hydroxide are effective for foot rot (Timmer 1977). Dipping of seedlings in captan solution before planting is recommended in U.S.A. (Grimm and Whidden 1966). Fumigation of nursery bed with methyl bromide has proved most effective to avoid nursery infection (Grimm and Alexander 1971). Leaf fall and fruit rot could be controlled by spraying bordeaux mixture (1%) or bordeaux mixture plus tin sulphate or difolatan (0.3%) (AICFIP 1982). In addition, spraying of Ridomil MZ and Aliette (0.2%) have also been reported effective (Rawal 1990). Diseased leaves and fruits may be collected and burnt or burried deeply.

Anthracnose : The disease is a serious problem and a limiting factor in U.S.A. and Carribean countries. In India, the disease has been reported from Punjab, Madhya Pradesh, Uttar Pradesh, Assam, Rajasthan and South India. The disease is also known as withertip which in recent years is responsible for decline in production of lime in North India, particularly in Kanpur area. Withertip, leaf-spot and fruit rot, all are caused by Colletotrichum gloeosporioides Penz. (Pathak 1980).

Symptoms : Anthracnose attacks the young leaves, shoots, blossoms and fruits of small acid limes. Young foliage and blossoms are blighted and distinct lesions formed on leaves and fruits. Affected fruits frequently drop prematurely. Withertip is characterized by shedding of leaves and die-back of twigs. Leaves show light green spots which turn brown. On dead twigs, black dot like acervuli appear in concentric rings. The stem end of immature fruits results in fruit drop. In severe cases, branches show die-back and the tree dies in a few years. Symptoms of anthracnose appear on leaves, young shoots and tender fruits. On leaves, the necrotic spots show acervuli arranged in concentric rings. Dead parts of the twigs assume silvery grey appearance. Twigs show a slight gumming and a sharp line of separation between healthy and dead tissues. Affected buds fail to develop and affected fruits drop off. Often, the infected fruits develop reddish brown stain on the rind. The fungus has also been held responsible for russeting and tearstaining of rind. This type of symptoms can also be caused by rust mites. The disease is also caused by Gloeosporium limetticolum Clousen or G. foliicolum Nishida. Malta oranges and grape fruits suffer much more damage. C. gloeosporioides remains in a dormant condition in the dead twigs and branches. This leads to the twigblight. The infected hyphae produced by appresoria remain latent even after the fruit mature and produce anthracnose in oranges and grape fruit, if the peal is injured or fruits are over matured.

Control Measures : The disease can be minimised through proper management of orchards by proper irrigation, manuring, cultural practices and plant protection measures. Dead twigs should be pruned and destroyed, cut ends should be protected by bordeaux paste. Such trees should be sprayed with bavistin (0.1%) or captafol (0.2%) 3 times after pruning (AICFIP 1987). Chowdhuri (1936) advocated the pruning of diseased twigs in winter followed by spraying with bordeaux mixture in February and March, and September.

Singh and Sinha (1954) reported that spraying the fruits with careful deposition of spray at the stem-end, using 4:4:50 bordeaux mixture, in 3:5 lime-sulphur, 0.1% copper sulphate solution and 2 per cent formalin, considerably reduced fruit drop in preliminary trials. Satisfactory control of anthracnose of lime has been obtained through spray of trees by bordeaux mixture (5:5:50) or 0.33 per cent perenox. Spraying of bavistin (0.1%) or Topsin-M(0.1%) at fortnightly intervals are also effective during fruiting period.

Powdery Mildew (Acrosporium tingitanium Carter Subr.)

The disease is common in India and is serious on mandarin and sweet oranges in the sub-mountain tracts of Coorg, Nilgiris, Pulney, Wynad and Shevaroy hills in South India and in parts of West Bengal, Assam and Sikkim in the northeastern India. The disease has been reported from Bangalore on lemons in addition to above group (Rawal and Ullasa 1988).

Symptoms : All the aerial parts of the plant are attacked. The disease is characterized by a whitish, mealy powdery growth on young leaves and twigs, which covers the entire leaf and young growing shoots. Even the small fruits also get infected. Such fruits shrink and become mummified. Affected leaves become distorted. The powdery coverage of foliage and twigs leads to leaf drop and die-back (Devarajan 1943). The affected fruits drop off prematurely and the yield is reduced.

In Coorg, the disease appears in a severe form between October and March. Damp mornings with a few hours of sunshine favour onset of the disease. The pathogen enters the epidermis and absorbs nutrients through haustoria. The mycelium covers the host surface. Conidia are disseminated by wind. Such conidia germinate ectophytic and later on become endophytic. After infection, the fungus leaves purple tinge on mature leaves.

Epidemiology : The disease is influenced by high humidity and defused light. The host parts which are devoid of surface free water get infected soon. Hence, cloudy and cool nights are most congenial for the disease to develop fast. The water shoots coming out under the tree canopy are the first to get infected.

Control Measures: The disease is easily controlled by dusting the foliage with finely powdered sulphur (Devarajan 1943; Ramakrishan 1954), and spraying of cosan, thiovit and calixin. Prophylactic spray should be at a proper time to achieve effective control.

Scab (Elsinoe faweetti Bitancourt)

It is a common disease of sour orange, Rough lemon and Tangelo, and has been reported from West Bengal, Assam, Uttar Pradesh, Andhra Pradesh and Karnataka and is mostly prevalent in the nursery. It is also recorded on sweet oranges in Karnataka (Rawal and Ullasa 1988). About 25-55 per cent of the fruits are completely spoiled in Assam (Chowdhuri 1955) by scab.

Symptoms: It rarely leads to the leaf drop but foliage distortion by the buff coloured warts may suggest thinness of the top, especialy in nursery. On leaves, small, semi-translucent dots appear usually on the underside of young leaf that become well defined pustules. The lesions mostly appear singly on one side of leaf. A few days later, these projections develop a cream to yellow-orange colour at the tips. The opposite surface corresponding to warty growth shows a circular depression with a pink to red depression. Severely attacked leaves become distorted and occasionally shed prematurely. The lesions may also appear on shoots. Sprouts of trees and the stem of nursery plants develop scab lesions similar to those on leaves.

Other parts of the tree on which scab lesions may be found are tender twigs, blossom pedicels and buttons. Eruptions occur either as individual round pimples or on the confluence of many infections. Heavily infected fruits may drop shortly after being attacked and those remaining on the trees may be scarred and distorted. Scab sometimes may be confused with similar effects produced by the other agencies. It does not invade the flesh. The lesions on fruits have corky projections which often break into scabs. When fully formed scab lesions are raised and are buff to pink to drab in colour (depending on the stage of infection).

Epidemiology : The fungus produces mostly the conidial stage (Sphaceloma fawcetti var. scabiosa Jenkins) on the host. The pathogen perpetuates and survives in off season as ascospores. Conidia are formed from 7-30°C with 66-100 per cent humidity. The fungus can infect tissue only when the surface is wet, but prefers 16-23°C temperature (Pathak 1980). Its spread is through rains. In India, the disease is a problem in areas with low temperature and high humidity. The water is the most important factors influencing severity of the disease (Yamada 1961) and can infect fruits in summer as well as in spring.

Host Resistance: The varieties/lines reported to be resistant against this disease are Sohymndong, Milan, Khatta Jamir, Brazillian Rough Lemon and Chase Rough Lemon (Rawal 1990).

Control Measures: Spraying of bordeaux mixture has been reported by Kar and Saha (1943), and Chowdhuri (1953). Spraying with difolatan (0.2%) also gives effective control (Rawal 1990). Ferbam 7.5 lb 75 WP or fixed copper have been recommended elsewhere. However, Ferbam has been found to be superior over copper spray.

Twig Blight (Colletotrichum gloeosporioides Penz.)

It is also characterized as die-back of twigs in large numbers preceding by the shedding of leaves and is present wherever citrus is grown. This disease is also caused by the association of Fusarium spp. and Diplodia natalensis Pole.

Symptoms : Blight is characterized by three types of symptoms. Wilting of foliage, die-back and post-emergence decline of watershoots. The first indication of blighted trees are dulling of the foliage followed by wilting. Symptoms may appear throughout the conopy. Gradually the leaves roll tightly, loose turgidity and drop. Die-back sets in after the leaves are shed. Sometimes the leaves shed due to the attack of powdery mildew and leaf rot pathogens. Such twigs also show die-back symptoms. Successive flushes in affected portions of the trees are retarded and dwarf or fail to develop. Such affected trees often bloom heavily, however, fruit set is poor.

Epidemiology : The disease has been found to be predisposed by the attack of powdery mildew and other leaf spot and rot causing agents. Adverse soil pH and deficiency of zinc, manganese and boron have also been suggested by Childs (1953). However, the C. gloeosporioides has also been seen to cause the vascular obstruction. Many times the insect infestation also led to the fungal infection. Though the disease appears in both spring and rainy season, it is more prevalant in the rainy season (Sohi and Kapoor 1990). The soil pulverization and adequate fertilization and maintenance of adequate soil moisture is very important. Spraying of proper insecticide is also recommended to control the insect attack. Pruning of the diseased twigs followed by the spraying of bavistin (0.1%) or captafol (0.2%) 3 times at monthly intervals have been found to be the most effective (Rawal 1990).

Alternaria Leaf Spot Fruit Rot (Alternaria citri Ellis & Pierce)

An important foliage disease in the nursery, particularly on Rough lemon and Rangpur lime.

Symptoms : Lesions develop following infection of young leaves and appear within the leaf blade or along the margins. They are regular to irregular spots surrounded by chlorotic halo, may coalesce to involve much of the leaf area. The colour of the necrotic tissue ranges from light to dark brown, edges being darker. Severe attacks lead to the leaf drop. This also attacks fruit. A decay of fruits on trees usually is visible externally, but sometimes the decay is observed when the fruits are cut opened. Juice of a few fruits with black rot may taint an entire box. The most conspicuous symptoms appear a few weeks after colour breaking stage. Affected fruits develop an orange colour prematurely. The infected fruits usually show light brown to blackish discolouration of the rind near the styler end. Internal symptoms consist of a brown to greenish black discolouration and disorganization of the core, more clear at the styler end. Even a little infection leads to the bitter flavour to the flesh. Affected fruits are ultimately shed.

Control Measures : Spraying of copper oxychloride fungicides (0.3%) or Dithane Z-78 or chlorothalonil (0.2%) give a very effective control. Infected fruits may be collected and destroyed.

There are other diseases, namely, wilt (Fusarium spp.), dry root rot (Macrophomina phaseolina), Diplodia gummosis (Diplodia netalensis), pink disease (Botryobasidium salmonicolor), malanose (Phomopsis citri) and other root rots due to Ganoderma lucidium and Armillariella mellea which frequently occur in the fields and need attention.

Citrus Canker : The disease was first recorded in Japan. Now it is known to occur in every citrus growing area of the world.

Symptoms : It may effect any part of the tree above the ground. The leaves, twigs, young branches and fruits are the most susceptible. It first appears as small watery, transluscent spots, usually dark green in colour than the surrounding tissue and with a raised convex surface. As disease advances, the surface of the spots becomes white or greyish and finally ruptures, exposing a light brown, spongy, central mass developed in crator like formation. The spots usually surrounded by a halo. On young twigs, lesions are similar to those on leaves and fruits but on older twigs they are irregular in shape. Fruit lesions are not surrounded by yellowish halo and the crater like appearance is more pronounced. In severe cases, fruits may crack and drop. Grape fruits, sweet oranges lime and lemon are very much susceptible to canker. Casual organism is Xanthomonas campestris pv vitri (Hasse) Dye.

Disease Cycle : The pathogen survives on diseased twigs and leaves. It enters the host through various types of wounds, natural openings, etc. On penetration in the host, it multiplies in the intercellular spaces, dissolves the middle lamella and establishes itself in the cortex. The disease is chiefly disseminated by wind, rain and insects. Citrus leaf miners are very important in the dissemination of this disease. However, man seems to be the chief agent of dissemination.

Control : It is mainly attained through: (1) quarantine, (2) eradication and destruction of infected trees, (3) spraying of neem cake at the rate of 7 kg per acre have been reported to be highly effective in checking citrus canker as well as leaf minor, (4) spraying of streptomycin sulphate (0.05%) at 75 days interval. Phytomycine (0.02%) has also been reported to be effective in checking the disease, and (5) spraying with copper oxychloride or bordeaux mixture (1%) along with pruning of infected plant parts were found effective.

Jackfruit

Jackfruit (Artocarpus heterophyllus) is consumed as fruit as well as vegetable. The trees are supposed to be hardy but even then the following diseases become a limiting factor in its cultivation.

Leaf Spot (Colletotrichum gloeosporioides)

It is most common disease occurring on Jackfruit. The disease is characterized as dark brown to brick red spot on both the leaf surfaces. These spots later turn into greyish white centres with dark brown boundaries. Dark coloured acervuli of the fungus are observed on greyish region which, when torn, ooze out profuse spore masses. The fungus has been seen to infect the young tender shoots and petioles of the leaves also. The disease is effectively checked through the sprays of carbendazim (0.1%) or methyl thiophanate (0.1%) or captafal (0.2%) or chlorothalonil (0.2%) (Rawal and Saxena 1997).

Die-back (Botryodiplodia theobromae)

It is a destructive disease of Jackfruit. The onset of die-back becomes evident by discolouration and darkening of the back some distance from the tip. The dark area advances and young green twigs start withering first at the base and then extending outwards along the veins of leaf edges. The affected leaves turn brown and their margins roll upwards. At this stage, the twigs or branches die, shrivel and fall, and there may be an exudation of gum from affected branches. Such branches have also been found to be affected by shoot borers and shot hole borers. The infected twigs show internal discolouration when split open. In early stages, epidermal and sub-epidermal cells of twigs are often slightly shrivelled. On such twigs, erumpent aceruvuli of Colletotrichum gloeosporioides are also observed. To check this disease, pruning of infected twigs followed by spraying of carbendazim (0.1%) or Topsin M (0.1%) or chlorothalonil (0.2%) have been found effective (Rawal and Saxena 1997). The control of shoot borers and shot hole borers by spraying suitable insecticides is also recommended.

Pink Disease (Botryobasidium solmonicolor)

It is widespread in tropical and subtropical areas. The disease appears as a pinkish powdery coating on the stem. Pink colour represents profuse conidial production of fungus. Young woody branches of the affected trees loose their leaves and show die-back. Similar disease has also been recorded on mango. The affected branches should be pruned and the cut end should be pasted with bordeaux paste or copper oxychloride (Pathak 1980).

The leaf spot due to Phyllosticta artocarpina has been reported but does not seems to be serious on Jackfruit.

Mangosteen

Garcinia mangostana L. is the popular fruit of South-East Asia. Though not much work has been done as regards the pathological problems occurring on mangosteen, however, a few diseases have been reported in the literature.

Thread boll caused by Pellixularia koleroga Cool has been reported to be the common disease in Puerto Rico. This fungus also goes to coffee, cocoa, rubber trees and many other plants and is common under high humid conditions. This disease first appears as a filmental growth on young stem which gradually goes on young leaves and starts attacking from the margins where it forms the white mat like structures. The immature fruits are susceptible to this disease which become mummified and are unfit for consumption.

The affected leaves begin to loose their luster, turn brown, dark brown and finally black leading to their fall. The leaves have a typical web like structures on them. The disease has been reported to be checked by spraying bordeaux mixture (5:5:30) or copper oxychlorides. Wilt (Zignoella garcineal P.) is a common disease in Malaysia which causes the formation of tubers on the new stems that are later seen as mature tubers of the older stem. This results in the wilting and death of the plant which affects first leaves, then branches and later the entire tree. The best method of control for this diseases is to cut and destroy such branches well in advance to avoid further spread of the disease.

Rambutan

Rambutan (Nephelium lappaceum L.) is one of the best known fruits of South-East Asia (Muchjajib 1990; Tindal 1993). The trees are severely attacked by powdery mildew caused by Oidium nephelii. It attacks leaves, flowers and young shoots. The affected flowers drop off leading to failure of the crop. The disease is favoured by the dry humid weather. The disease can be best checked by spraying wettable sulphur, bayleton or systhane fungicides.

Fruit rot due to Phytophthora necotianae var. parasitica and P. botryosa are recorded to be serious in most areas. Anthracnose (Colletotrichum sp.) causes severe damage on leaves, inflorescence and young fruits. Stem canker (Fames lignosus) has been reported to be serious in the Philippines. Phomopsis leaf spot, stem canker (Dolabra sp.) and stem die-back (Thyronectrichia pseudotria) are recorded to be the minor diseases. Then there are post-harvest diseases caused by various fungi like Aspergillus spp., Botryodiplodia theobromae Port, Colletotrichum spp., Fusarium spp., Pestalotia spp. and Phomopsis homopsis. Most of these are also pathogenic in the field.

Litchi

Litchi chinensis Sonn. as such does not have much of a disease problem (Galan Sanco 1989, 1991). However, tar leaf spots caused by Pestalotia pauciseta Sacc., Botryodiplodia theobromae Pat. and Colletotrichum gloeosporioides occur in litchi orchards in some areas. The leaves occasionally showed mixed infections of B. theobromae, and C. gloeosporioides and some spots may reveal the presence of P. pauciseta. The lesions coalesced at the later stages. Older leaves show higher percentage of infections than the younger leaves. Generally, the incubation period of P. pauciseta, B. theobromae and C. gloeosporioides ranges from 5-10 day, 3-7 days and 4-8 days respectively. The diseased leaves serve as the source of primary infection. The leaf spots spread rapidly during humid and rainy months. The fungicides like bavistin, Topsin M, baycor and dithiocarbmates will control the disease.

There is another serious leaf spot of litchi incited by Microdiplodia litchi Pathak and Desai that has been recorded at Udaipur, Rajasthan. The diseased leaves show yellow brown to brick red areas mostly towards margins. The coloured areas gradually become light brown and show black dot like pycnidia. The other problems on litchi as listed are due to Clitocybe tabescence. Sudden wilt caused by Fusarium sp., die-back due to Phomopsis and foliar necrosis by Gloeosporium spp. are reported. There are some of the post-harvest diseases which are carried along with the fruits during transit and storage. Among them, the pathogens like B. theobromae, C. gloeosporioides, Cylindrocorpon forkinense, and Phomopsis spp. and Pestalotia species are the most important and need attention as they are causing diseases on the plants also.

5

Pruning Fruit Trees

Introduction

While the principles of pruning fruit trees do not change, the actual practices used in modern production systems vary. The higher density-supported training systems now used by commercial growers are managed by the same principles of pruning used in the past. Before embarking on a specific training system for high-density plantings, investigate specific techniques required for training and pruning that system.

Principles

Pruning Dwarfs Trees

Numerous experiments show that pruning is a dwarfing process, that is, it reduces the total size of the tree. Leaves are the food-manufacturing organs of the tree. Cutting away a live branch, which if left would have borne leaves, reduces the output of the "factory", and the final result is a reduced bearing area. Summer pruning has the greatest dwarfing effect.

Pruning Appears to Invigorate Trees

There is a certain deception in the effects of pruning, in that strong shoots with large leaves tend to rise just at the back of pruning cuts. This gives the impression of increased growth. Pruning reduces the number of growing points, thus stimulating an increase of growth at the remaining points. However, pruning, in proportion to its severity, reduces the total growth and the total leaf surface of the tree. (Invigorate trees when necessary with appropriate fertilizer use.)

Pruning Effect is Localized

The removal or cutting-back of the laterals of a branch reduces the growth of that branch. Pruning has the direct effect of producing growth response in the immediate area in which the cut is made.

Pruning Too Heavily Has Several Ill Effects

Excessive pruning, by over-stimulating growth, causes loss of fruit colour, delayed fruit maturity, and growth of suckers and watersprouts. Excessive succulent growth increases the hazard of fire blight in apple and pear, canker in peach, and winter injury in all species.

Pruning the Young Tree Delays Fruit Bearing

Because pruning tends to force the growth of long, succulent shoots that grow late in the season, manufactured food is not allowed to accumulate in sufficient quantities to cause the formation of fruit buds. Consequently, the tree is kept in a juvenile or nonfruiting condition for a greater number of years. Young trees often grow very upright, and become so dense that the grower is tempted to thin-out the branches. It has been shown, however, that early fruit bearing will open out a tree much more effectively than pruning. With the young tree, remember: light pruning, early bearing, a spreading tree.

Clean, Flush Cuts Heal More Quickly

Make pruning wounds flush with the limbs to which the unwanted branches are attached. The exception to this rule is with peach. With peach use a collar cut rather than flush cuts. Wood healing is more rapid when the bark ridge at the base of larger limbs is not removed during a pruning cut. Where one-year-old wood is being pruned, make the cut as close as possible to a bud to facilitate healing. A stub and/ or ragged edges at these points greatly delays the healing of the wound and increases the probability of drying out and infection. This is particularly important with peach because canker may enter where healing is delayed.

Narrowed-Angled Crotches are Weak

Where the crotch angle is less than 35 degrees the attachment will be weak because of the inclusion of bark. The tissues in narrow crotches are slower to mature in the fall and may be injured by low temperatures, especially in test winters. Narrow crotches are usually further weakened by water, ice, rot organisms and canker. Thus, remove limbs that make sharp angles early. This avoids possible loss of a large part of the tree later on due to breakage from weight of fruit.

Pruning and Training at Planting Time

When new fruit trees are dug in the nursery, a large percentage of the finer root system is left behind during the process. The new tree that once had a balance of leaves to roots now develops more leaves than there is root system to sustain. The resulting tree will be out of balance, resulting in poor growth. To overcome this problem, heavily prune all fruit trees at planting time before growth starts. Remove at least ýÿ of the potential leaf area. Eliminate all branches below 60 cm. If the tree is tall enough, cut back the leader to about 80-90 cm, and remove all shoots. With well developed trees or older trees having some limbs you wish to retain, you can cut these limbs back to 2 or 3 buds and retain growth in that area. Totally remove limbs:

- with poor crotch angles
- that are too low to the ground or
- large limbs approaching the calliper of the central leader.

Consider the pruning at planting time and over the next 2 or 3 years as a training process. The final strength of the tree depends upon the wise selection of branches and your ability to maintain the proper balance between these branches. Mistakes made in this formative period may mean weak trees and, in addition, the correction of errors may call for severe pruning in later years-the removal of considerable portions of the bearing area and the creation of large wounds subject to infection. It is critically important that you build a strong framework in the early years.

The central leader training system is recommended for all fruit trees. This tree "Christmas tree" form is conical in shape, with a wide base and a narrow top. During the early life of the fruit tree, keep this tree with judicious pruning. Some trees, particularly sour cherry, peach and Japanese plum, do not retain a dominant central leader for long. This is not a problem. The central leader tree at maturity consists of 6-8 main scaffold branches distributed vertically and spirally around the trunk, and with the topmost branch (leader) well in the lead of the lower ones.

The vertical distance between limbs occupying the same quadrant of the tree will vary from 10-80 cm depending on the ultimate size of the tree. Limbs too close together will result in excessive shading. This weakens the limb and leads to poor fruit quality, reduced productivity and ultimately failure of the shaded limb.

If the branch angles at the main trunk are sufficiently wide (over 35 degrees), there will be no bark inclusions in the crotches, resulting in a strong tree.

With many cultivars the central leader will slow in growth, making removal or heading unnecessary. Keeping the central leader in the early years encourages wider angles on the framework branches below it. The central leader of dwarfed trees will be lost prematurely if allowed to fruit too soon.

Training the Young Nonbearing Tree

The less pruning during this period, the more quickly the tree comes into bearing. Consequently, once the main branches are selected, do minimal pruning until the tree is bearing. An early crop of fruit, besides bringing early financial returns, slows down vegetative growth and bends down the branches. This aids in controlling the height of the tree, and opens up the tree to permit more light and better spray coverage. Guard against too-early fruiting of the central leader.

Prune lightly in the third and fourth years, chiefly by thinning-out rather than heading-back. Generally avoid heading-back from the second year, until the trees are in heavy bearing.

Remove branches that form narrow crotches (less than 35 degrees) with the trunk. Eliminate branches that grow straight up or into the tree, those that are weak and drooping, and those that tend to cross or otherwise interfere with others.

Six or 8 main branches are usually sufficient to build a good tree. With pear varieties susceptible to fire blight, and where fire blight is likely to be serious, prune especially lightly and leave more framework and secondary branches.

Pruning the Mature Fruit Tree

Dormant Pruning

Most pruning is done when the trees are dormant, between the time when the leaves drop in late fall and when the buds begin to swell in early spring. The safest and best time is just before the buds swell. The most risky time is very late fall and early winter. Dormant pruning of peach and nectarine increases the risk of winter injury; prune during the bloom period.

In the orchard, start spring pruning early enough to be completed before the leaves appear. The risk of winter injury increases if pruning is begun too early. Pruning followed by low temperatures means winter injury-not always seen but almost sure to be present. The amount of injury is directly related to the length of time between the pruning operation and the temperature drop; the shorter the time, the greater the injury.

All pruning has a dwarfing effect, but dormant pruning produces the most new growth. If you want new vegetative growth, dormant pruning is the way to get it. The harder the cutting, the greater is the response in new shoot growth. The response takes place in the area of the tree where the cuts are made.

Pruning During Bloom

Most growers prefer not to prune peach and nectarine trees until the flower buds have advanced sufficiently to assess the flower winter survival. Delaying pruning much beyond shuck split may cause a serious loss of tree vigour.

Early Summer Pruning

Pruning has the greatest dwarfing effect in June and early July. If you wish to reduce vegetative growth and prevent shoots developing, this is the time to prune.

But remember that early-summer pruning has a very dwarfing effect. It first dwarfs the root system, and then the whole tree.

Midsummer Pruning

Pruning at this time of year has little or no effect on stimulating new vegetative growth. At the same time, it is not nearly so dwarfing as early-summer pruning. The root system is dwarfed somewhat but only moderately, as compared with the results of early-summer pruning. This may be the time to reduce the height and width of your trees by cutting back the new growth. The amount of cutback would depend on the growth, vigour and age of the tree. A well-grown tree with a good crop could have new growth reduced by 1/2 to 2/3. This will let in more direct sunlight to colour the fruit. It will also reduce vegetative growth, which makes more sugars available to the developing fruit, and results in improved flavour, although fruit size may be reduced.

Fall Pruning

Pruning cuts will not heal at this time of year, but in order to spread the work load over more time, some pruning might be started in early fall. Start with the oldest trees, and cease all pruning operations well before any chance of a severe temperature drop. Do not prune peach and nectarine trees in the fall because of the ever-present threat of canker.

Summary of Rules for Pruning Bearing Trees

1. Cut out broken, dead, or diseased branches.
2. Where 2 branches closely parallel or overhang each other, remove the least desirable, taking into account horizontal and vertical spacing.
3. Where possible, prune on the horizontal plane; that is leave those laterals on the main branches that grow horizontally or nearly so, and remove those that hang down or grow upward.
4. Thin all varieties to permit thorough spraying and the entrance of sunlight and air.
5. Where it is desired to reduce the height of tall trees, cut the leader branches back moderately to a well-developed horizontal lateral.
6. Prune the lower branches of broadheaded or drooping varieties to ascending laterals.
7. Varieties that tend to produce numerous twiggy, lateral growth should have some of this growth removed to prevent overcrowding.
8. Make close, clean cuts. Stubs encourage decay and canker, thus forming a source of injury to the parent branch or trunk.

9. Prune moderately. Very heavy pruning is likely to upset the balance between wood growth and fruitfulness, and generally should be avoided.
10. Prune regularly. Trees that are given some attention each spring are more easily kept in good condition than trees that are pruned irregularly.
11. Prune that part of the tree where more growth is required. This is particularly important with old trees. New growth will be stimulated only in those parts of the tree that were pruned. Reduce pruning to an absolute minimum where growth is already excessive.

Do not remove a branch unless there is a very good reason for doing so. Leaves are the food-manufacturing organs, and if the leaf area is reduced unnecessarily, the tree will be reduced in growth or fruitfulness or both.

Fruit Tree Pruning

Pruning fruit trees is a technique that is employed by gardeners to control growth, remove dead or diseased wood or stimulate the formation of flowers and fruit buds. The most economical pruning is done early in the season, when buds begin to break, and one can pinch off the soft tissue with one's fingers (hence the expression "nipped in the bud"). Many home fruit growers make the mistake of planting a tree, then neglecting it until it begins to bear. But careful attention to pruning and training young trees will ultimately determine their productivity and longevity. Good pruning and training will also prevent later injury from weak crotches that break under snow or fruit load.

Overview

To obtain a better understanding of how to prune plants properly, it is useful to have some underlying knowledge of how pruning works, and how it affects the way in which plants grow.

Plants form new tissue in an area called the meristem, located near the tips of roots and shoots, where active cell division takes place. Meristem growth is aimed at ensuring that leaves are quickly elevated into sunlight, and that roots are able to penetrate deeply into the soil. Once adequate height and length is achieved by the stems and roots, they will begin to thicken to give support to the plant. On the shoots, these growing tips of the plant are known as apical buds. The apical meristem (or tip) produces the growth hormone auxin, which not only promotes cell division, but also diffuses downwards and inhibits the development of lateral bud growth which would otherwise compete with the apical tip for light and nutrients. Removing the apical tip and its suppressive hormone allows the lower dormant lateral buds to develop, and the buds between the leaf stalk and stem produce new shoots which compete to become the lead growth. Manipulating this natural response to damage (known as the principle of apical dominance) by processes such as pruning (as well as coppicing and pollarding) allows the horticulturist to determine the shape, size and productivity of many fruiting trees and bushes. The main aim when pruning fruit trees is usually to obtain a decent crop of fruit rather than a tree with an abundance of lush yet unproductive foliage.

Unpruned trees tend to produce large crops of small, worthless fruit often damaged by pests and diseases, and much of the crop is out of reach at the top of the tree. Branches can become broken by the weight of the crop, and the cropping may become biennial (that is, only bearing fruit every other year). Overpruned trees on the other hand tend to produce light crops of large, flavourless fruit that does not store well. Pruning is therefore carried out to achieve a balance between shoot growth and fruit production.

Formative Pruning of Bush Trees

Formative pruning of apple and pear trees (the pome fruits; the stone fruits such as cherries, plums, gages, etc., have different requirements and should not be pruned during the dormant months) should be carried out during the dormant winter months between November and March (or June and September in the southern hemisphere) and during the early years of the tree's life to develop a strong framework capable of bearing the weight of the crops that will be borne in later years. This involves hard pruning, although in later years pruning will be lighter and carried out to encourage fruiting.

Maiden Tree

A maiden whip (that is, a one year old tree with no side shoots) should be pruned to a bud with two buds below it at about 80 cm from the ground immediately after planting to produce primary branches during the first growing season. A feathered maiden (that is, a one year old tree with several side branches) should have its main stem pruned back to three or four strong shoots at 80 cm from the ground. Side shoots should be shortened by two thirds of their length to an upward or outward facing bud. Lower shoots should be removed flush with the stem.

Two Year

Remove any lower shoots and prune between three and five of the best placed shoots by half to an upwards or outwards facing bud to form what will become the tree's main structural branches. Remove any inwards facing shoots.

Three Year

Prune the leading shoots of branches selected to extend the framework by half to a bud facing in the desired direction. Select four good laterals to fill the framework and shorten these by a half. Prune any remaining laterals to four buds to form fruiting spurs.

Four Year

The tree will have begun to fruit and only limited formative pruning is now required. Shorten leaders by one third and prune laterals not required to extend the framework to four buds.

Five year and Onwards

The tree is considered to be established and should be annually pruned as described.

Pruning the Cropping Tree

Before pruning it is important to distinguish between spur bearing and tip bearing varieties. The former, which is the most common type, bear most of their fruit on older wood, and include apples such as *Cox's Orange Pippin, James Grieve* and *Sunset,* and pears such as *Conference, Doyenne du Commice* and *Williams Bon Chretien.* Tip bearers on the other hand produce most of their fruit buds at the tips of slender shoots grown the previous summer, and include the apples *Worcester Pearmain* and *Irish Peach,* and the pears such as *Jargonelle* and *Josephine de Malines.* There are basically three types of pruning that are applied once the main shape of the tree has been established.

Renewal Pruning

- Spur pruning:Spur bearing varieties form spurs naturally, but spur growth can also be induced.
- Renewal pruning: This also depends on the tendency of many apple and pear trees to form flower buds on unpruned two year old laterals. It is a technique best utilised for the strong laterals on the outer part of the tree where there is room for such growth. Pruning long neglected fruit trees is a task that should be undertaken over a lengthy period, with not more than one third of the branches that require removal being taken each year.
- Regulatory pruning: This is carried out on the tree as a whole, and is aimed at keeping the tree and its environment healthy, eg, by keeping the centre open so that air can circulate, removing dead or diseased wood, preventing branches from becoming over crowded (branches should be roughly 50 cm apart and spurs not less than 25 cm apart along the branch framework), and preventing any branches from crossing.

Pruning of tip Bearers

Tip bearers should be pruned lightly in winter using the regulatory system. any maiden shoots less than 25 cm in length should be left untouched as they have fruit buds at their tips. Longer shoots are spur pruned to prevent over-crowding and to stimulate the production of more short tip bearing shoots the following year. Branch leaders are 'tipped', removing the top three or four buds to a bud facing in the desired direction to make them branch out and so produce more tip bearing shoots.

Fruit Tree Propagation

Fruit tree propagation is usually carried out through asexual reproduction by grafting or budding the desired variety onto a suitable rootstock.

Perennial plants can be propagated either by sexual or vegetative means. Sexual reproduction occurs when male pollen from one tree fertilises the ovules (incipient seeds) of the flower of another, stimulating the development of fruit. In turn this fruit contains a seed or seeds which, when germinated, will become a new specimen. However, the new tree will inherit many of the characteristics of both its parents, and it will not grow 'true' to the variety from which it came. That is, it will be a fresh individual with many unpredictable characteristics of its own. Although this is desirable in terms of increasing biodiversity and the richness of the gene pool (such sexual recombination is the source of most new cultivars), only rarely will such fruit trees be directly useful or attractive to the tastes of humankind. A tendency to revert to a wild-like state is common.

Therefore, from the orchard grower or gardener's point of view, it is preferable to propagate fruit cultivars vegetatively in order to ensure reliability. This involves taking a cutting (or scion) of wood from a desirable parent tree which is then grown on to produce a new plant or 'clone' of the original. In effect this means that the original Bramley apple tree, for example, was a successful variety grown from a pip, but that every Bramley since then has been propagated by taking cuttings of living matter from that tree, or one of its descendants.

Methods

The essentials of our present methods of propagating of fruit trees date from pre-Classical times. Grafting as a technique was first developed in China from where it was imported to Greece and Rome. Classical authors wrote extensively about the technical skills of fruit cultivation, including grafting techniques and rootstock selection. The oldest surviving named varieties of fruits date from classical times.

The simplest method of propagating a tree asexually is rooting. A cutting (a piece of the parent plant) is cut and stuck into soil. Artificial rooting hormones are sometimes used to assure success. If the cutting does not die of desiccation first, roots grow from the buried portion of the cutting to become a complete plant. Though this works well for some plants (such as figs and olives), most fruit trees are unsuited to this method.

Root cuttings (pieces of root induced to grow a new trunk) are used with some kinds of plants. This method also is suitable only for some plants.

A refinement on rooting is layering. This is rooting a piece of a wood that is still attached to its parent and continues to receive nourishment from it. The new plant is severed only after it has successfully grown roots. Layering is the technique most used for propagation of clonal apple rootstocks.

The most common method of propagating fruit trees, suitable for nearly all species, is grafting onto rootstocks. These are varieties selected for characteristics such as their vigour of growth, hardiness, soil tolerance, and compatibility with the desired variety that will form the aerial part of the plant (called the scion). For example, grape rootstocks descended from North American grapes allow European grapes to be grown in areas infested with *Phylloxera,* a soil-dwelling insect that attacks and kills European grapes when grown on their own roots. Grafting is the process of joining these two varieties, ensuring maximum contact between the cambium tissue (that is, the layer of growing plant material just below the bark) of each so that they grow together successfully. Two of the most common grafting techniques are 'whip and tongue', carried out in spring as the sap rises, and 'budding', which is performed around July and August.

Bud Grafting

1. Cut a slice of bud and bark from the parent tree.
2. Cut a similar sliver off the rootstock, making a little lip at the base to slot the scion into.
3. Join the two together and bind.
4. In time, the scion bud will grow into a shoot, which will develop into the desired tree.

Apple Rootstocks

Another reason for grafting onto rootstocks is that this enables the grower to determine the tree's eventual size. Apple tree size classes number one to ten in increasing height and breadth. A "1" is a dwarf which can be productive and as short as three (3) feet with proper pruning. A "10" is the standard sized tree with no dwarfing and will grow to twenty(20) or more feet tall and wide, dependent upon the variety chosen. In general the class range is (1) 10-20% of full size, (2) 20-30%, (3) 30-40% and so forth to size 10 which is 100% of full size.

Apple tree rootstocks are referred to by numbers prefixed by letters indicating the developer of the rootstock.

"M" designates Malling series developed stocks. East Malling Research is a pioneer in the development of dwarfing rootstocks. East Malling Research Station in Kent, England collected clones of the Paradise stocks from France in 1912 from which 24 "M" were designated with no particular order to the rootstock characteristics other than where they were located in the garden at the time the numbers were assigned. In other words, M.2 is larger tree than M.9 while M.27 is smaller than M.26.

"MM" designates Malling-Merton stocks developed from joint breeding program by John Innes Institute, in Merton, England, & East Malling Research Station in the early 1950s.

The "MM" series was developed primarily to provide resistance to Woolly Apple Aphid (Eriosomatinae) infestation.

"EMLA" designates East Malling/Long Ashton research stations who took the "M" stocks and developed virus free versions. E.g., EMLA 7 is M 7 with a guaranteed virus free stock. EMLA characteristics are often different from the parent "M" rootstock. Note that nearly all the apple rootstocks in the industry are now virus free.

"CG" or "G" designates Cornell-Geneva stocks which are those developed via the Cornell & USDA collaboration at the New York Agricultural Experiment Station in Geneva, NY. The "G" is the old designation. All newer stocks are "CG" followed by numbers that actually provide some information about the stock. As one might surmise, this is a huge improvement in the classical naming scheme which has no identification method at all.

Diagram illustrating comparative sizes of apple trees depending on rootstock grafting

- M.27 Malling 27: A very dwarfing rootstock. Unless the central leader is supported, the tree will be very small. Often only used as an intermediate stem piece on MM.106 or MM.111. If handled and spaced properly, it can be a very productive stock for a vertical axe system. Trees can be grown three to four feet tall and produce about 45 fruit, roughly 2 pecks, depending on fruit cultivar.
- M.9: Very dwarfing-Reaches a height of 8 to 10 ft (2.4 to 3.0 m), coming into fruit after 3–4 years, reaching full capacity of 50 to 65 pounds (23 to 29 kg) after 5 to 6 years. It will grow under average soil conditions, but needs a good rich soil to thrive. A good choice where space is limited and fertility is high. Permanent staking is required, as is routine feeding and watering. Trees on this rootstock always require leader support. The rootstock is very susceptible to fire blight and can develop burr knots.
- G.41 Geneva 41, released in 2005, produces trees the size of M.9. The rootstock was developed from a cross between M.27 and Robusta 5 made in 1975. Resistant to Crown | Collar | Root rot(Phytopthora) and fire blight.
- M.26: Dwarfing-Similar to M9 in effect, although somewhat more vigorous and generally stronger, with a higher expected eventual yield of 65–75 pounds (29–34 kg) and height of 8 to 10 ft (2.4 to 3.0 m). A good choice where soil quality is average and compact growth is required. Comes into fruit after 3-4 years, reaching full cropping capacity after 5 to 6 years. Staking needed for first five years of its life. It is susceptible to collar rot and fire blight and should

not be planted in a wet site. Certain varieties when grafted onto this rootstock may exhibit signs of graft union incompatibility(i.e., the union breaks).

- G.11 Geneva 11 is the second release of the Cornell breeding program similar in size to M.26 but more productive. Has the advantage of being resistant to fire blight and crown rot as well as only rarely producing suckers or burr knots.
- G.202 Geneva 202(CG 5202) is a semidwarfing rootstock that produces a tree in class 5 slightly larger than M.26 and is more productive than M.26. It was developed from a cross of M.27 and Robusta 5 to be fire blight and Phytopthora resistant as well as having resistance to woolly apple aphids. In a 9-year study with the scion cultivar of the "Liberty" apple, G.202 was about 50 percent smaller than M.7 but had much greater production efficiency.
- M.7 Malling 7 rootstock produces a semidwarf tree of Class 6 that is freestanding in deep well drained soils but in rocky, steep, or shallow soils, it tends to lean. The rootstock may sucker profusely and is susceptible to collar rot(Phytopthora).
- MM.106: Semi-dwarfing-Sometimes referred to as semi-vigorous, this is the most widely used of rootstocks. It is probably the best choice for the average garden under average conditions, being tolerant of a wide range of soils, and producing a tree with an eventual size of 14 to 18 ft (4.3 to 5.5 m). Trees on this stock begin producing fruit within three to four years, and yield 90 to 110 pounds (41 to 50 kg) after some seven or eight years. MM106 is very suitable for use with weaker varieties that would produce under sized bushes with more dwarfing rootstocks. Can be trained as a half standard tree, but is rather too vigorous for cordons unless the soil is poor. Requires staking for the first five years or so of its life. Trees on MM.106 are highly susceptible to collar rot especially when planted in soils that remain wet(poor percolation).
- M.111 : Vigorous-Not generally suitable for garden scale growing, being both too large and spreading (18-25'), and too slow to come into cropping. They are however suitable for growing as specimen standards in the large garden, or for producing medium sized bushes on poorer soils. Begins to fruit after six or seven years, reaching full capacity of 160 to 360 lb (73 to 160 kg) after eight to nine years. It is the most cold-hardy rootstock readily available. Planting depth of this rootstock is critical. The union should be no higher than 1 to 2 inches above the final soil line.
- M.25: Very vigorous-Suitable for a grassed orchard, and to grow on as a full standard. Plant 20 ft (6.1 m) apart, makes a tree of 15 to 20 ft (4.6 to 6.1 m) or more height and spread, eventually yielding 200 to 400 lb (91 to 180 kg) per tree. This rootstock is primarily used in UK and is rarely seen in USA where M.111 is used for this size tree.

- Seedling: Very vigorous trees produced on a rootstock grown from seed. There is greater variability than with the vegetatively propagated rootstocks. Apples used for production of seedling rootstocks include 'Dolgo' and 'Antonovka', which are both extremely hardy and vigorous.

That is only a sample of some of the more important current apple rootstocks that are available. There are at least a hundred more that have been developed to either provide enhancement or prevent potential damage from one kind of pest or another.

The problem with growing fruit trees, especially apple trees, is that they are subject to many different types of damage from bacteria, fungi and insects. The general approach of the commercial industry has been to use as many chemicals as necessary to insure attractive and marketable fruit. The attitude, still prevalent, has been "Who cares? Nobody eats a tree!" but as environmental problems increase and the general public pushes for low or no-spray fruit, there has become a commercial need for fruit that does not require such intensive spray programs. This is being achieved, albeit slowly, by rootstocks and trees that are bred to have natural disease and pest resistance.

The Malling series and clones have been standard rootstocks for apples for many years and remain the standard "workhorses" for the commercial industry (in USA). However, since most of them are susceptible to disease some Malling rootstocks are being replaced by new breeds, including the Cornell-Geneva series. One of the newest rootstocks, only released commercially in 2004, is CG5202 (G.202) which adds resistance to the woolly apple aphid(WAA) for the "CG" series of stocks which already has resistance to the major problems preventing quality production of apples utilizing organic control systems. Combined with highly resistant trees such as "Liberty" it is showing great potential.

That leads to another characteristic of rootstocks that is or can be bred into them: environmental adaptability. This may be tolerance to wet | dry soil conditions, acidity | alkalinity of soil or even hot | cold air temperature.

Some new rootstocks based on Siberian Crab Apple are being used in colder areas for more cold tolerance.

The ability of rootstocks to modify or augment characteristics of fruit trees is limited and often disappointing in the final results. It takes ten years to get a full picture of the effects of any one rootstock so a rootstock that appears promising in the first five years of a trial may fail in the last five years. The Mark (apple) rootstock was such a stock and has now fallen mostly into disfavour. Another, the G.30, has proved to be an excellent stock for production but it was only after a number of years of trials that it was found to be somewhat incompatible with "Gala" apple(and possibly others) so that it is now recommended to be staked and wired.

To get a clear picture and push the industry forward, a consortium was founded and the so-called "NC-140" trials of rootstocks began. These test many pome rootstocks in many different sites across the USA and thereby provide growers, be they backyard or commercial, a clearer picture of what to expect when growing fruit trees on specific stock, in specific planting methods in their specific area of the USA. As one can imagine, this has the potential for a large economic benefit to both growers and consumers as well as going a long way to eliminating the need to spray pesticides as frequently as is currently required.

Pear Rootstocks

Pears are usually grafted onto quince rootstocks, which produce small to medium sized trees. Some varieties however are not compatible with quince, and these require double working. This means that a piece of pear graft-work compatible with both the quince rootstock and the pear variety is used as an intermediate between the two. If this is not done the pear and the rootstock could eventually separate at the graft. Varieties that require double working include 'Bristol Cross', 'Dr Jules Guyot', 'Doyenné d' été' and 'Williams Bon Chrétien'.

- Quince C: Moderately vigorous- Makes a bush pear tree about 8 to 18 ft (2.4 to 5.5 m) tall, bearing fruit within four to eight years. Suitable for highly fertile soils and vigorous varieties, but not where conditions are poor. Used for bush, cordon and espalier growing. Old stocks of Quince C may be infected with a virus, so care should be taken to obtain certified virus free stock. If in doubt, use Quince An as there is not a great amount of difference in vigour between the two.
- Quince A: Medium vigour- Slightly more vigorous than Quince C, this is the most common variety upon which pears are grafted. Bears fruit between four to eight years, making a tree of some 10 to 20 ft (3.0 to 6.1 m) in height and spread. Suitable for all forms of pear trees except standards.

Pear stock: Very vigorous- Pears grafted onto pear rootstocks make very large standard trees, not suitable for most gardens.

Cherries

Until the 1970s, cherries were grown of the vigorous Malling F12/1, Mazzard (*Prunus avium*), or Maheleb rootstocks, which required much space and time before cropping began, thus the growing of cherries was not a realistic option on a garden scale.

The introduction of the rootstock 'Colt' enables trees reaching a maximum height of 12 to 15 ft (3.7 to 4.6 m) to be grown, and if trained as a pyramid it is possible to restrict growth to about 10 ft (3.0 m). The popular sweet variety 'Stella' could even be grafted onto a 'Colt' rootstock and successfully grown in a pot on the patio.

Plums

Plum rootstocks include;

- Pixy- A dwarfing rootstock, suitable for bush trees planted 8-10 (3 m) apart.
- St. Julien A- A semi vigorous rootstock suitable for bush and half standards planted 12 to 15 ft (3.7 to 4.6 m) apart. Also suitable for peaches, nectarines and apricots.
- Brompton or Myrobalan B- Suitable for half standards planted 18 to 22 ft (5.5 to 6.7 m) apart. Also suitable for peaches, nectarines and apricots.
- Myro-29C- Semi-dwarf rootstock. Shallow, vigorous, good choice for hard soils. Somewhat drought tolerant.
- Citation- Semi-dwarf rootstock. Shallow, vigorous, good choice for hard soils. Prefers a wetter soil.

Own-Root Fruit Trees

Some species of fruit are commonly grown on their own roots; new plants are propagated by rooting, layering, or modern tissue-culture techniques. In these cases there are may be no great advantages to using a special rootstock or improved rootstocks are not available. Fig, filbert, olive, pomegranate, gooseberry, bramble, and other fruits are commonly grown without any special rootstock.

Pruning and Training Thornless Blackberries

Thornless Blackberries: For two years after planting, thornless blackberry primocanes tend to grow along the ground, like a vine. Growers may have to move trailing canes in the direction of the row to allow room for cultivation. Generally, only a small crop of fruit is produced in the first season. If growth is poor during the first season, cut the canes back to several inches in late winter to force development of sturdier, more fruitful canes.

In the second and succeeding years, shoot growth is more vigorous and upright. Tie these new shoots to the trellis when they reach a length of four to six feet. Some growers prefer to wait until after harvest to remove old canes before tying new shoots to the trellis in a fan shape (do not bunch them). In the spring before growth starts, prune any laterals back to twelve inches to encourage larger fruit.

Thornless blackberries have been grown successfully using a variety of trellising systems which are required to hold canes above the ground. If the tops of new canes are pruned during the summer to keep growth below three to four feet, no trellis is needed. Construct the blackberry trellis by stretching a wire between posts set twenty feet apart in the row. For erect blackberries, use one wire attached to the post about thirty inches from the ground. For semitrailing blackberries, use two wires at heights of three feet and five feet from the ground.

Erect blackberries such as Cherokee and Cheyenne require pruning out of the root suckers that arise from the crown. During the growing season, it is desirable to allow root suckers to develop to about a twelve-inch-wide row. Any growth beyond this should be eliminated.

When the new shoots of erect blackberries reach thirty to thirty-six inches in height, cut off the tips. This will force branching lower on the canes and will cause the canes to thicken, making them better able to support a heavy fruit crop. During the winter, prune the laterals to twelve to fourteen inches for convenient harvesting and larger berries. In late winter, remove any remaining dead or weak wood. Leave healthy, vigorous canes spaced at six canes per linear foot.

Pruning the old canes is critical to the prevention of disease. As soon as the last fruit has been picked in summer, cut all the old canes and burn them. Also, prune damaged or weak canes, leaving four to eight new shoots. Some growers tip primocanes when they reach a height of twenty-four inches so that low growing laterals are more easily protected during winter.

Managing Fruit Trees in the Home Landscape

A summary of management requirements for the fruits and nuts most commonly grown in Texas.

Propagation

One of the most common questions among home gardeners is how various fruit and nut crops are propagated. The most popular methods of propagation for Texas crops.

Plant Nutrition

Under natural conditions, plant nutrition is usually not a problem because slow growth and low production are acceptable. But the homeowner usually wants many large fruits as soon as possible, so a judicious fertility program is necessary.

When buying a fertilizer, it is important to know what the numbers mean. A fertilizer label usually has three numbers, such as 10-10-10. These numbers represent the percentages of nitrogen, phosphorus and potassium, respectively, in the mix. In other words, a 50-pound bag of 10-10-10 has 5 pounds each of N, P_2O_5 and K_2O. But that accounts for only 15 pounds out of a 50-pound bag. The rest is either other salts or inert fillers, such as sand, perlite or other materials.

Nitrogen. Nitrogen is necessary for all vegetative growth-roots, leaves, stems, flowers and fruit. Nitrogen deficiency causes the plant's lower leaves to turn yellow, while excessive amounts delay maturity, causing increased vegetative growth and decreasing cold-hardiness. Nitrogen is needed in all parts of Texas to maintain healthy plants.

Phosphorus. A long-standing fertility recommendation has been to apply a high-phosphorus fertilizer anytime a plant fails to bear. However, in most situations, phosphorus does not increase blooming in a fruiting plant. Phosphorus fertilization is needed most in sandy soils, since clays and loams ordinarily have adequate amounts for perennial plants.

Iron. Many areas have problems with iron deficiency even though adequate amounts are present in most Texas soils. The alkaline (pH over 7.0) soils tie up iron, making it unavailable to the plants. Iron deficiency is normally found in new growth and is characterized by yellow leaves with a "road map" appearance because almost all the veins remain green.

Zinc. Zinc deficiencies are a problem in pecans in most of the state. Zinc deficiencies are characterized by small leaves, rosetted (highly branched, short, bunchy) growth, twig and limb dieback and sometimes tree death. Leaf veins have a band of green on each side, giving the leaves a striped appearance.

Choosing A Fertilizer

Because Texas has such variability in its soil types, a single broad fertilizer recommendation is not applicable in all cases. The best means of determining your individual soil need is to have a soil test done by the Texas Agricultural Extension Service's Soil Testing Laboratory. Information on sampling techniques and prices is available from your county Extension office. Where soil tests are unavailable and a general recommendation is needed, use the following guidelines. On sandy soils, use a complete fertilizer such as 15-5-10 for most fruit crops. Acceptable rates are 1 pound per inch of trunk diameter per year of age for trees up to 10 years old and 1 pound per 10 feet of row for blackberries and grapes.

On loams and on clay soils, the only element generally required is nitrogen, which can be furnished with 1 pound of 21-0-0 per year of age or inch of trunk diameter up to 10 for trees or 1 pound per 100 feet of row for vine crops. Iron deficiencies are common on most of the alkaline soils in Central, South and West Texas. Generally foliar iron sprays do not work well to correct iron deficiencies in fruit crops. Soil-applied iron chelates (FeEDDHA) at label rates are the most effective. Iron sulfate (copperas) is seldom effective as a remedy for iron chlorosis of fruit trees or berries. Where iron deficiencies are a problem, do not use fertilizers containing phosphorus, as they usually make the problem worse.

Zinc deficiency on pecans is a problem in most of Texas; therefore, make at least three applications of a foliar zinc spray each year during April and May for maximum growth. Do not spray other plants with zinc at the rates used on pecans since leaf burn and defoliation may occur. Zinc nitrate is least likely to cause burn to other plants in the landscape. When zinc deficiency is a problem on other crops, it can usually be corrected by soil applications of zinc sulfate.

Water Requirements

To maximize production and maintain plant health, it is important to irrigate plants through periods of stress. There are many ways of determining when and how much to irrigate. Begin watering when the soil begins to dry out after a rain. If your area experiences prolonged wet periods, wait about 2 weeks before starting.

It is important that you apply enough water to wet the soil to a depth of at least 12 inches with each irrigation. You can check the irrigation depth by pushing a metal rod into the soil at the wetted area. Keep pushing until resistance stops the rod's penetration: the length of the underground portion of the rod is the irrigation depth. When the soil is wetted to a depth of 12 to 18 inches at each watering, plants usually do not suffer from drought stress, even when waterings are missed. In areas where the lawn is not watered, trees still require water. You can provide water in a number of ways, but trickle or drip irrigation is one of the best methods. Trickly irrigation kits for lawn and garden use are readily available. When using drip irrigation, place enough emitters around the plants to water 50 to 70 percent of the root system. During hotter months of the growing season, operate drip systems a minimum of 10 to 12 hours, 1 day per week, to maintain adequate soil moisture. They can be controlled by electric time clock, by switching tensiometer or by hand.

Training and Pruning

Training and pruning are essential parts of the management program for any fruit or nut crop. Training the plants helps to ensure that they will grow properly and bear successfully-and that they will be attractive additions to your landscape.

Tree Training Systems. The two most commonly used tree training systems are the open centre and the central leader systems. Open centre systems are generally used on peaches, plums, apricots and almonds. Apples, pears, pecans and persimmons are normally trained to a central leader. There are good and bad points to both systems, but neither system dictates that the trees be pruned a certain way. In fact, many trees are not pruned or trained at all. However, contrary to some popular beliefs, pruning is important for producing most fruit and nut crops. But don't get bogged down in specifics; instead, keep general principles in mind. Both training systems start out in essentially the same way. Fruit trees are 30 to 36 inches tall or larger, or nut trees 4 to 7 feet tall, are cut back by one half at planting with all side shoots removed. This forces out strong vigorous shoots which can be easily trained to the desired system.

The open centre or vase system of training simply involves maintaining a framework of branches around an open "vase" in the middle of the tree. This allows sunlight to penetrate into all parts of the tree, allowing for good production in all areas. The key to open system training is to develop a strong open centre framework in the first 2 or 3 years and to subsequently maintain this shape. This later pruning involves the heading-back of shoot terminals to outward-growing branches, the removal of large, fast-growing branches that fill the open centre and the removal of crowded branches and any diseased or broken limbs. This reduces the height and keeps the centre of the tree open.

The central leader system consists of a central trunk around which scaffolds (primarily side branches) of the desired number and spacing can be arranged with wide-angle crotches. Three to eight scaffold branches are commonly developed from the central leader trunk.

A "modified" central leader tree is cut back each winter and a new central leader shoot is selected each spring. Pecans, apples and pears are generally pruned in this manner. The top centre of modified central leader trees is often thinned out for better light penetration into the interior of the tree canopy. Uniformly space the scaffolds around the central leader.

Further information for pruning specific crops is available in these Texas Agricultural Extension Service publications:

- B-1591, "Home Fruit Production-Figs."
- B-1598, "Home Fruit Production-Pears."
- B-1607, "Home Fruit Production-Apples."
- B-1629, "Home Fruit Production-Citrus."

Grape Training. Grapes require severe pruning to develop high shoot vigour during training and to maintain production of quality berries on mature vines. Prune back leaving only two buds at planting.

First growing season: Allow growth to develop at random to establish a good root system.

First winter: Prune off all growth except one shoot with two good buds.

Second growing season: Choose the most vigorous shoot and train it up a stake. Tie the vine to the stake every 6 inches. Keep side shoots pinched off, but keep leaves on the trunk. When the shoot reaches just above the cordon wire, pinch out the tip to force lateral branching. Train laterals (arms or cordons) down the wires, tying regularly to keep the cordon straight and in place.

Mature vines: Cane pruning and cordon pruning are two basic systems commonly used. With either system, winter-prune the vine, leaving a total of about eight buds of the previous summer's growth prior to the third year and 20 to 40 buds per vine on mature vines.

Cane pruning is best for varieties that produce few fruit on basal buds such as Thompson seedless and for small clustered varieties. With this system, all four arms are removed each winter.

One-year shoots (renewal canes) are tied to the wires to replace the old arms. Cut off the tip of each renewal cane at a point where it is approximately 3/8 inch in diameter.

Cordon pruning is most commonly used in Texas vineyards. This system consists of leaving about seven upright spurs (one to two buds per spur) on each cordon. All other growth is removed. Use a "clothesline-like" T-top trellis with a wire at each end to more efficiently catch and spread the upright can growth. For this method, a 2-foot wire T-top replaces the single top wire shown.

Muscadine grapes are usually cordon-trained, but because of greater vine vigour, muscadine vines can be allowed to support up to four cordons with spurs of three to five buds each.

Grape arbors are easiest to maintain using cordon training. Develop parallel, spur-pruned cordons about 2 feet apart across the top of the arbor. Black Spanish and Champanel are excellent arbor varieties. Muscadines also make good arbor grapes in East Texas.

Berry Training. The types of blackberries and raspberries range from erect, freestanding plants to trailing vines. The training system differ according to the type of growth.

Erect blackberries and raspberries produce low, sprawling growth the first year after planting, but in the second and subsequent years new growth is very erect. Clip the tips from new canes two to three times from May to September to force side-branching and to develop a full, compact hedgerow. Prune the hedgerow much as you would an ornamental hedge. A well-pruned hedgerow of erect berries does not need trellis support.

Fruiting canes die soon after the fruit have matured. An accumulation of dead canes poses a considerable nuisance when picking berries, so it is best to keep dead canes removed.

Winter pruning is not needed if the berries are properly pruned in the summer.

Trailing and semierect blackberries and raspberries should be trellised. A proper trellis allows the canes to be spread for good sunlight exposure.

Tip new canes two or three times each summer to encourage more branched growth. Leave new canes (primocanes) of trailing blackberries on the ground each winter to help protect them from freeze damage. Trellis the canes before they begin growth in mid-March. Remove and destroy fruiting canes (floricanes) after harvest in the summer as soon as they die.

6

Gelified Sugar Fruit Preserves

Technology of Fruit Jellies

Jellies are gelified products obtained by boiling fruit juices with sugar, with or without the addition of pectin and food acids. Jellies are usually manufactured from juices obtained from a single fruit species only, obtained by boiling in order to extract as much soluble pectin as possible.

Jellies have to be clear, shiny, transparent and with a colour specific to the fruit from which they are obtained. Once the product is removed from the glass receptacles where it was packed, jellies must keep their shape and gelification and not flow, without being sticky or of a too hard consistency.

Technological flow-sheet for jellies manufacturing covers two categories of operations: those to obtain gelifying juices and those related to the manufacturing of jelly itself.

Production of Gelifying Juices

Washing & Sorting are carried out in usual conditions;

Cutting is applied eventually only to pomaces (apples, quinces) and are limited to cutting in halves or quarters;

Boiling Is Performed With Water Addition With 50-100% water, needed for pectin extraction. The boiling time is 30-60 min., it should not be longer so as to avoid pectin degradation; at the same time boiling must not be too violent.

Juice Separation is carried out by a simple drain through metallic sieve or cloths; in these cases the yield is lower and the residue can be used for marmalade production. In bigger productions it juice separation by hydraulic press is preferred, yield being in these cases greater. Juice Clarification is strictly necessary in order to obtain clear jellies. This step can be achieved by sedimentation during 24 hours or by filtration.

Manufacturing of Jellies

Basic Recipe Setting is done starting with equal parts in weight of sugar and juice (for example 1000 g juice and 1000 g sugar). As final jelly has to contain about 60% added sugar, weight of finished product must be of about 1600 g, by evaporation of about 400 g water.

Boiling is carried out as following: juice is boiled up to removal of about half of the water that has to be evaporated, then the calculated sugar quantity is added gradually; the remainder of the water is evaporated until a concentration in soluble substances (refractometric extract) of 65-67% is reached, in which is incorporated also the sugar from juice.

During boiling it is necessary to remove foam / scum formed. Product acidity must be brought to about 1% (malic acid) corresponding to pH > 3. Any acid addition is performed always at the end of boiling.

For juices rich in pectin, gelification will occur without pectin addition. If at the trial boiling test the gelification has not occurred, because of pectin absence, in this case 1-2% powder pectin will be added by operating as indicated: pectin is mixed with 10-20 fold sugar quantity and is introduced directly in the partially evaporated juice and then boiling is conducted rapidly up to final point. Evaluation of final point is done not only by refractometry but also by gelifying test.

A rapid test for evaluation of juice pectin content is possible by mixing a small sample of juice with an equal volume of 96% alcohol; the apparition of a compact gelatinous precipitate indicates a sufficient pectin content for gelification.

Boiling of jellies is performed in small batches (25-75 kg) in order to avoid excessively long boiling time which brings about pectin degradation.

Cooling is optional and is carried out up to 85 °C, in double wall baths with water circulation.

Filling is performed at a temperature not below 85 °C in receptacles (glass jars, etc.), which must be maintained still about 24 hours to allow cooling and product gelification.

Receptacle Closing is done after product gelification.

Usual jelly types are: quinces, strawberries, cherries, wild berries, alone or in mixes with apples.

Grading of Marmalades

Three categories can be defined:

- fine marmalade, manufactured from one fruit;
- superior marmalade, obtained from a mix of fruit in which 30% are "noble" species (cherries, strawberries, apricots, etc.) and 70% from other species;

- marmalade from fruit mixes; apples, pears, plums, quinces, ungrafted apricots and wax cherries may be used, with the optional addition of "superior" fruit which was rejected at sorting but which was sound.

The content in total soluble substances (refractometric extract) of marmalades must be 64% minimum; the acidity must be between 0.5% and 1.8% expressed as malic acid.

Basic Recipe Setting for a normal composition - marmalade without pectin addition the following is a basic recipe:

100 kg semi-processed fruit product:

(10% refractometric extract) ...10 kg soluble substances

55 kg sugar ....55 kg soluble substances

155 kg 65 kg soluble substances

55 kg water to be evaporated

100 kg marmalade with 65% refractometric extract.

This marmalade satisfies many standards and at same time has a good shelf-life since it contains less than 35% water. Semi-processed fruit products must have a minimum 8% refractometric extract; in this case the recipe should use 125 kg of raw material, with 80 kg water to be evaporate.

The use of semi-processed fruit products with a low refractometric extract presents the following drawbacks:

a. higher water quantity to be evaporated;
b. longer boiling times with negative impact of pectin degradation;
c. loss of flavour and
d. lower equipment efficiency.

Pectin addition in marmalade manufacture produces the following advantages:

a. improvement of gelification,
b. economy in fruit;
c. shorter boiling time; this maintains taste and flavour and produces higher equipment efficiency.

Pectin addition makes it possible to obtain the "fine" type of marmalade from "noble" fruits which do not contain enough pectin (cherries, peaches, apricots, etc.). In marmalades from fruit mixes, low pectin content can be compensated by addition of semi-processed fruit products which are rich in this component (for example apples).

When pectin is to be added, the above recipe should be modified as follows:

80 kg semi-processed fruit product

(10% refractometric extract) ...8 kg soluble substances

55 kg sugar ....55 kg soluble substances

10 kg pectic extract (10 % R.E.) ......1 kg soluble substances

145 kg 64 kg soluble substances

45 kg water to be evaporated

100 kg marmalade with 64% refractometric extract

Pectin can be added as pectic extract with about 10% refractometric extract (R.E.) in the recommended proportion or in the form of a powder considered with 100% dry matter (*e.g.* 100% soluble substances) in a quantity of about 1%.

In some countries it is usual to add 5-12% of corn syrup (calculated to finished product weight) to replace the corresponding quantity of sugar (100 parts corn syrup can replace 80 parts sugar). Corn syrup has to be liquefied by heating before use. Corn syrup addition reduces the too sweet taste of marmalade, avoids sugar crystallisation and gives a special shine to finished products.

Marmalade manufacturing follows the technological line drawing and covers the following steps:

"Mark" Preparation can be achieved from fresh fruits (as indicated in the section related to semi-processed fruit products) or starting from chemical preserved semiprocessed fruit products: "marks" or "pulps". In the latter case, pulps will be processed in marks which then will be desulphitated.

Desulphitation is carried out by boiling at atmospheric pressure, under vacuum in specialised equipment or under pressure in special retorts built in acid-resistant material. In any case, the desulphitation must be carried out before sugar addition because sugar will bound to the sulphur dioxide. The desulphitation operation must be conducted so as to be, if possible, fully completed; the finished product must contain less than 0.005% free SO_2.

The technological flow for marmalade production according to the following: fresh fruit after sorting on control belt (1) is washed in a washing machine (2) is brought to the continuous boiling equipment (3) then to the pulper (4); the semi-finished mark is passed on to the storage tank (6). Pulps are boiled and desulphited in continuous boiling equipment (3), then are brought to pulper (4) and to the storage tank (6).

Boiling aims at evaporating the required water quantity, to facilitate the formation of pectin-sugar-acid gel and to partially invert sugar (about 40% from total sugar). The boiling operation can be carried out in open kettles or in evaporators under vacuum. In the latter case, the warm "mark" from storage tank (6) is aspirated in a concentrator (7) in a vacuum and submitted to a partial boiling up to removal of half of water quantity which needs to be evaporated; the calculated sugar quantity is then added by aspiration, keeping the boiling on.

After this the pectin extract or powder pectin which has previously been dissolved in warm water, is added; when the final concentration is reached, as indicated by refractometric control, the required quantity of acid is added. Sugar is added in proportion of 55% in finished product, pectic extract (10% refractometric extract) at a level of about 10-15% and the acid (citric, tartaric, lactic) in a quantity needed to obtain a finished product acidity of about 1%.

Boiling at atmospheric pressure affects not only the appearance but also the nutritional value of the products, mainly if these contain proteins, as some albuminoids coagulate even at 60°C.

Food products for which flavour is an essential property as for example fruit juices, etc., are also affected by the action of heat. Heat treatment has an impact on vitamin losses, mainly of vitamin C, in the presence of oxygen as is the case at concentration in open vessels.

Technological Line for Marmalade Production

Sugars are generally less damaged by heat at temperatures below 100° C; as the boiling point is increasing above 100° C, a risk of partial sugar caramelization exists.

In the study of heat effects on products submitted to concentration operation, it is necessary to take into account not only the evaporation surface temperature but also the distribution of the temperature in the whole liquid mass.

The length of the heating period also has a major influence because in many cases it is preferable to concentrate the liquid at a relatively high temperature in a short time avoiding the drawbacks of lower temperatures acting during a long time.

In order to maintain the food value and organoleptic properties, it is necessary that concentration take place at a low temperature which can be achieved by concentration under partial vacuum, taking into account that boiling point decreases when the residual pressure decreases, respectively with the increase of vacuum degree.

Advantage of concentration under partial vacuum are the following:

- lowering of boiling point;
- the total time needed for concentration of food products under a residual pressure of about 200 mm Hg is about half as compared to the that of concentration by boiling at atmospheric pressure;
- by lowering the concentration temperature and time, organoleptic properties and of nutritional value are maintained better particularly as far as the vitamins are concerned;

- when products are concentrated in a vacuum, it is possible to recover volatile aromatic substances by using adequate installations.

Technical procedures of concentration by vaporisation can be classified in:

a) concentration at atmospheric pressure: continuous or discontinuous;

b) concentration under partial vacuum: discontinuous (in vacuum equipment with simple or multiple effect) or continuous (in vacuum installations with continuous action or in thin film vaporisation installations).

Even if open kettle equipment is less expensive than evaporators in a vacuum, it is necessary to take into account that boiling under vacuum has the following advantages:

a) low boiling temperature (60-70° C), depending the degree of the vacuum; this give the fruit better taste and flavour-keeping qualities;

b) easy feeding with raw and auxiliary materials;

c) shorter boiling time;

d) better working conditions (vapour elimination in condensed water and not in open air).

There are small size evaporators under vacuum which can be well suited to the needs of medium size operations in developing countries.

Cooling of marmalade to about 50-60°C can even be done in a vacuum evaporator by closing the heating steam and maintaining vacuum degree or by discharge in storage tanks.

Filling in receptacles (boxes, jars, glasses, etc.) is done preferably with filling machines followed by labelling. Small packages can be closed warm or after complete cooling; big packages (boxes, etc.) must be closed only after cooling, *e.g.* 24 hours after processing.

Storage of marmalade must be done in dry rooms (air relative humidity at about 75%), well ventilated, medium cool places (temperature 10-20 degrees C), disinfected and away from direct sunlight and heat. These measures are necessary because marmalade is a hydroscopic product and, by water absorption, favourable conditions for mould development are created.

Technology of Special Fruit Jams

Special fruit jams are products similar to marmalades but in which fruit partially keep their shape (whole, halves, etc.). Special fruit jams are obtained by boiling fruit with sugar, with or without pectin addition, with acid addition followed by n concentration by evaporation.

Special fruit jams present a pronounced gelification at their cooling and can be considered as fruits included in a pectin-sugar-acid gel.

High quality special fruit jams are obtained only from fresh fruit or possibly frozen and from only one fruit species.

Special fruit jams are classified in:

a) non-pasteurized (min. 68% refractometric extract) and

b) pasteurized (min. 65% refractometric extract); minimum acidity, expressed in malic acid, is 0.5%.

Basic Recipe Setting is done taking into account following elements: - maintenance, as much as possible, of fruit shape, because this is specific to these finished products;

- added sugar, in relation to finished product, must be at 60-65%; this high percentage is needed in order to obtain a rapid and strong gelification and to facilitate conservation of fruit shape;
- a satisfactory gelification cannot generally be obtained without pectin addition, at a level of 1-2%, or respectively 10-20% pectic extract (10% refractometric extract);
- partial (about 40%) inversion of added sugar is necessary.

The basic recipe is: 80 kg fruit + 100 kg sugar + 1.6-2.0 kg pectin powder and 1 kg citric or tartric acid; this will yield about 165 kg of special fruit jam with 60% added sugar; water quantity to be evaporated is about 18 kg. If the finished product has to contain 65 % added sugar, the boiling has to be continued up to evaporation of about 30 kg water, the resulting finished product quantity will be about 153 kg.

Technological flow sheet for manufacturing of special fruit jams is as follows:

Fruit Preparation: sorting, washing, peeling and coring (for apples, pears, quinces), or removal of quetches and stones/pits (for plums, peaches, apricots, cherries) or of short tails (for strawberries and wild berries). Pomace fruits are then cut in slices or quarters.

Boiling With Sugar is the most important operation in production of special fruit jams and has as objective to evaporate water until gel formation and partial inversion of sugar. Boiling has to be conducted in such a way as to avoid fruit disintegration, but fruits must be well penetrated with sugar.

By boiling an equilibrium is reached, by osmosis, between sugar solution and cellular juice. The initial concentration gap between sugar syrup and cellular juice is very high and if the equilibration process is forced, juice comes out of cells and fruit loses its shape and may even disintegrate.

The boiling process accelerates the equilibration, intensity of which increases with temperature and boiling time.

Pectin addition shortens boiling and thus delays the equilibration; for this reason there are different methods for special fruit jams preparation:

- maintaining fruit in sugar, at ambient temperature, over 8-24 hours; fruit to sugar ratio is that indicated in recipe. After this sugar impregnation, fruits and resulting sugar syrup are brought together to boil; toward the end of boiling, pectin, dissolved in warm water, and necessary acid quantity are added.
- preparation of a very concentrated sugar syrup (at least 75%), in which fruit is introduced in order to be boiled.

The rest of the operations are similar to those described in the previous method

Soft fruit (strawberries, wild berries) can be mixed with sugar directly in evaporating open kettles, without added water and then heated gradually up to boiling, which is continued as in previous method.

Boiling is preferably carried out in small open kettles (50-100 kg) in order to avoid too long a boiling and fruit disintegration.

Gelification corresponds generally to the reaching of a concentration of 65% soluble extract, respectively 68% refractometric extract. Practical test for gelification is done as for jellies and marmalades.

Cooling is a technological step strictly necessary in order to avoid fruit rising to the surface in preservation receptacles. Cooling is done in a double bottomed water bath in which water circulates at about 80° C.

Filling of receptacles (jars, boxes, glasses, etc.) is carried out and it is necessary to assure at this stage that the finished product is homogeneous (equal quantities of fruits and gel).

Pasteurization is only applied to special fruit jams with 65 % refractometric extract packed in jars or boxes and is performed at about 100° C for about 20 min.

Gelification is carried out during product cooling and intensifies during storage.

Storage must follow the same conditions as for marmalades.

Non-gelified Sugar Fruit Preserves

Technology of Fruit Jams

Fruit jams are products obtained by boiling of fruits (whole, halves, etc.) with sugar syrup until the reaching a viscous consistency. Jams can be defined as fruits included in a concentrated syrup.

Jams are only prepared from fresh fruits; the usual product range covers the following species: strawberries, cherries, apricots, wild berries, peaches, plums, raisins, quinces, rose petals, etc.; manufacturing of jams from fruit mixes is not an accepted practice. Standards indicate that fruit jams have to conform with following conditions:

- fruit or fruit pieces content 45-55%;
- soluble substances, refractometric degree min. 72%;

- acidity (expressed in malic acid) min. 0.7%.

Basic Recipe Setting is presented below as an indication:

100 kg fruits

150 kg sugar

------

250 kg

35 kg water to be evaporated

------

215 kg jam at about 72% refractometric extract

Fruit Preparation is similar to special fruit jams with the difference that stone/ seed removal is mandatory for all species.

Boiling fruits with sugar can be achieved by the same three methods described above for special fruit jams. By boiling a sugar content equilibration is foreseen and the operation must be conducted in such a way that texture, flavour and colour of fruits be preserved. Foam/ scum has to be removed during boiling.

Generally boiling in concentrated sugar syrup (at least 75%) after a previous diffusion during 2-4 hours is in practical use. Boiling must to be carried out in small portions (about 15 kg) in order to avoid fruit disintegration.

For some fruit species, boiling has to be conducted in many phases / steps with "stops" to enable sugar diffusion. At the end of boiling, vanillin at a ratio of 125 g for 100 kg jam may be added for some fruits (white cherries, raisins, etc.). At the same time, it is also possible to add citric or tartric acid in order to avoid the "sugaring" defect.

Cooling of jams, necessary in order to avoid fruit rising to the surface, is carried out as for special fruit jams. Filling of receptacles and Storage for jams are performed in same way as for special fruit jams.

Special Non-gelified Fruit Marmalades

Special non gelified fruit marmalades are products resulting from fruit without stones or seeds, sieved or squashed, concentrated by boiling, without sugar added and non gelified. Their consistency results from a low water content (about 35%) and a high percentage of insoluble substances (5-10%). Sugar the from fruits acts as a preserving agent.

Plum special non gelified fruit marmalade is the product representative of this category. Other fruit is used very rarely, because they have a reduced sugar content as compared to plums; though there are some countries producing special non gelified fruit marmalade from pears or sweet apples.

For plums, the finished product in this category must contain minimum 55-60% soluble substances (refractometric extract), rising up to 70% for a high quality product.

Basic Recipe Setting for plums is done in relation with sugar content of fruits; when this is higher, the product quality is better and the yield is higher. Thus, if plums with 18% refractometric extract are used, 300-350 kg of fruits are needed for 100 kg finished product.

At the same time, concentration of sugar by boiling, also increases by three fold acidity and astringency of this special plum non gelified marmalade; for this reason it is recommended to use only sweet and completely mature plums.

Washing is Performed in Usual Conditions

Pre-boiling of fruits can be carried out in water or vapour, preferably with continuous running and has as its objective the softening of tissues.

De-stoning is Performed in a Pulper

Boiling of the sieved mass is done in double bottom open kettles with a big evaporation surface or in vacuum evaporators.

Boiling in open kettles enables production of a more tasty slightly caramelized, product; boiling in vacuum evaporators has the technological advantages indicated in marmalade production.

At the end of boiling and once of necessary concentration is reached, the product is poured directly into receptacles (drums, etc.) and left to cool in order to form a hard surface layer (crust).

Storage of well closed receptacles is carried out as for marmalade. Special non gelified fruit marmalade can also be prepared from chemically preserved semi-processed fruit products, but the quality is lower than that obtained from fresh fruits.

Sometimes dried prunes in a mix with preserved semi-processed products can be used for plum special non gelified marmalade preparation. In some countries plum finished products in this category are sweetened by the addition of maximum 30% sugar, calculated in relation to the finished product.

Technology of Fruit Pastes

These products are obtained in a similar way to marmalades and special non gelified marmalades, but have a lower water content (about 25%). Reduction of water content can be achieved by continuing the boiling of the product or by natural or artificial drying. A typical example of fruit paste without sugar added is the apricot paste - "pistil", etc. which is a concentrated special non gelified fruit marmalade poured in thin layers and sun dried. An example of fruit paste with sugar added is quince paste which is a marmalade concentrated by evaporation. Sugar content must be 65%; soluble substances content, 7075% refractometric extract and acidity at least

0.5% expressed as malic acid. Packing is done usually in polyethylene sheets and then in boxes or tins; storage conditions are similar to those for marmalade.

Technology of Fruit Syrups

Fruit syrups are products obtained by dissolving sugar in juices obtained from direct pressing of fruits. Sugar dissolving can be done at room temperature or by heating.

Syrups have to contain 68% soluble substances (refractometric extract) and minimum 1 g/100 ml malic acid. Up to a maximum 10% of sugar can be replaced by corn syrup. Syrups must be manufactured from the juice of only one fruit species.

Juice Preparation is carried out at room temperature as described above.

Sugar Dissolving In Fruit Juice can be performed by one of the following methods:

(a) `Boiling in open kettles, done by using the basic recipe:

350-400 kg fruit juice;

650-660 kg sugar;

max. 10 kg citric or tartric acid.

The juice is brought to boiling and the sugar is dissolved; the total time has to be as short as possible in order to avoid flavour loss and a too high sugar inversion degree (optimum inversion degree is 40%). Acid is added preferably towards the end of boiling.

During all boiling processes it is necessary to remove foam / scum. In order to avoid caramelization, the syrup has to be cooled rapidly, and this can be carried out in baths with double bottoms through which are circulated water.

One alternative to this method is to boil syrup in closed vessels to avoid flavour losses.

(b) *Boiling in a Vacuum:* The basic recipe is the same as above. Sugar and fruit juice are mixed previously in a pre-heating kettle and then transported to vacuum equipment.

Boiling is performed at 50° C and at the end the temperature is raised slowly up to 65-70° C. The syrup can be cooled directly in vacuum equipment by closing the steam inlet and by increasing the vacuum. In this boiling method it is possible to incorporate a flavour recuperation device.

(c) A continuous process for syrup preparation can be carried out by dissolving components with heat while passing them through a horizontal cylinder with a screw inside.

In the methods where sugar is dissolved by heat, it is also possible to use chemically preserved juices.

In this case it is necessary to first perform the desulphitation of juices preserved with SO2. This can be performed by boiling juice with optional water addition (and before any sugar addition). High quality syrups are obtained however from fresh juices.

(d) Sugar can also be dissolved at room temperature by using continuous flow percolators. These are similar to those used for salt solution preparation in vegetable canning processes. The juice goes over a sugar layer and is concentrated progressively until saturation (about 65 %). The syrup is then passed through a filtration section in the bottom of the percolator.

Syrup Filtration is needed in order to clarify crystals; the filtration of syrup is done in warm conditions through cloth.

Filling of syrup in bottles is done in aseptic conditions as much as possible in order to avoid syrup infection with osmophile yeasts. Syrup preservation is assured by the high sugar content with respect to a low water activity unclear.

Storage takes place in well ventilated storage rooms; avoiding sunlight at 10-15° C. The usual product range is: strawberries, cherries, wild berries, citrus fruits.

Technology of "Fruit in Syrup" Products

This type of product is represented by fruit (whole, halves or pieces) covered by a sugar solution and preserved by pasteurization. In these products sugar does not have a preservation effect but only a sweetening role.

Technological general flow-sheet covers the following steps:

Sorting is necessary in order to choose mature fruit, whole, unblemished, undamaged and with a specific texture. Fruit of good quality but with a texture not compatible with this type of finished product are used for the production of semi-processed products, marmalades, etc. This step is done manually on a sorting belt.

Washing is performed in equipment with fan and sprays. Cleaning covers removal of leaves, etc., skin removal by one of the described methods for some fruits and coring / pitting for others.

Cutting is applied only to pomace fruits and is performed preferably by mechanical means.

Preliminary Heat Treatment depends very much upon the fruit types and can be as different as a light blanching up to a real boiling; this step is aimed toward softening of tissues for hard fruits, elimination of waxy pruin layer (plums, etc.) or enzyme inactivation for pomace fruits.

This treatment has to be reduced to the minimum necessary in order to avoid sugar losses. Some fruit (for example apricots, black cherries, grapes, etc.) does not undergo preliminary heat treatment.

Cooling is carried out in water (as cold as possible) and should not be too long to avoid soluble substances loss.

Receptacle Filling Fruit is introduced manually or sometimes mechanically (filling tables, etc.) in receptacles and then sugar syrup is added.

Sugar Syrup Preparation is performed by dissolving crystal sugar in hot water (90100° C). After the sugar has been dissolved, the syrup is boiled briefly; removal of impurities and coagulated substances is then performed by foaming/scumming. An addition of about 0.3% citric acid helps syrup clarification, followed by filtration through cloth.

Preheating/Emptying of open receptacles has as its objective to eliminate air from fruit tissues and this step is largely replaced today by receptacle closing under vacuum. Preheating may be done in steam or in water in such a way that syrup temperature reaches 8090° C and is maintained for 10-15 min.

Hermetic Sealing is performed according to the type of receptacle and of cover/cap.

Pasteurization is carried out at 100° C and may be performed in continuous installations or in normal water baths. In some countries fruits in syrup are pasteurized at 80-90° C; this contributes to a greater flavour retention.

Cooling should be intensive in order to avoid discolorations and colour modifications and to reduce corrosion in metal cans.

Refractometric extract is to be measured 30 days after manufacturing. In order to establish exact concentration of sugar syrup to be added in each individual case it is possible to use the following formula:

$C = 1.5 \times [(2 \times E) - F]$

C = added syrup concentration;

E = refractometric extract needed in finished products, according to Table.

F = fruit refractometric extract.

Hot syrup is put into receptacles with a minimum temperature of 80° C in order to enable a corresponding vacuum after cooling.

It is possible to add in the syrup 0.3-0.5% citric or tartaric acid for fruit with a low acidity. Adding ascorbic acid in proportion of 0.8% assures colour maintenance and taste improvement based on its reducing action.

Avoidance of excessively soft texture in finished products may be achieved for soft fruit (strawberries, apricots, etc.) by dipping fruit in a solution of calcium chloride (5% $CaCl_2$ in water).

Finished product defects and production "accidents" for "Fruit in syrup" and the means of preventing them

- Change of colour in red fruit (red cherries, plums, etc.) to violet; to avoid this it is necessary to use glass jars or varnished/lacquered tinplate;
- Colour change to red for pears and quinces (pink shade) could arise in products which are over-pasteurized and insufficiently cooled;
- Change of colour of whole peaches peeled by chemical means is due to an incomplete inactivation of oxidative enzymes; to avoid this it is necessary to boil fruit after chemical peeling followed by washing with 0.5% citric or tartric acid solutions or by adding ascorbic acid in syrup;
- Softening of strawberries tissues can be prevented by a pre-treatment with warm syrup and also by addition of calcium salt.

Some technical data for processing are seen in Table.

Technical Data for "Fruits in Syrup"

Type of fruit/product		Minimum fruit	Soluble substances (in syrup), RE** at 20°C,
name	Size/shape	content, %	minimum
Apricots	Halves	55	25
Strawberries	Whole	47	25
Cherries	Whole	53	22
Quinces	Slices, 30 mm	50	26
Apples	Halves or quarters	47	25
Pears	Halves or quarters	57	22
Plums	Whole	45	25
Melon	Pieces	55	26
Wax cherries	Whole	53	27
Wild cherries	Whole	47	27
Pineapple	Slices	55	26
Mangos	Pieces	50	25
Papaya	Pieces	55	27
Fruit mixes	According to the	50	25
	conditions indicated		
	above, for each fruit		

** RE = Refract metric extract

Fruit Juice Technologies

Fruit juices are products for direct consumption and are obtained by the extraction of cellular juice from fruit, this operation can be done by pressing or by diffusion.

For the purpose of this document, the technology of fruit juice processing will cover two finished product categories:

- juices without pulp ("clarified" or "not clarified");
- juices with pulp ("nectars").

We will also define as :

a. "natural juices" products obtained from one fruit; and

b. "mixed juices" products obtained from the mix of two or three juices from different fruit species or by adding sugar.

Juices obtained by removal of a major part of their water content by vacuum evaporation or fractional freezing will be defined as "concentrated juices".

Technological Steps for Processing of Fruit Juices without Pulp

Fruit juices must be prepared from sound, mature fruit only. Soft fruit varieties such as grapes, tomatoes and peaches should only be transported in clean boxes which are free from mould and bits of rotten fruit.

Washing: Fruit must be thoroughly washed. Generally, fruit will be submitted to a pre-washing before sorting and a washing step just after sorting.

Sorting: Removal of partially or completely decayed fruit is the most important operation in the preparation of fruit for production of first quality fruit juices; sorting is carried out on moving inspection belts or sorting tables.

Crushing/Grinding/Disintegration Step is applied in different ways and depends on fruit types:

Crushing for grapes and berries;

Grinding for apples, pears;

Disintegration for tomatoes, peaches, mangoes, apricots.

This processing step will need specific equipment which differs from one type of operation to another.

Enzyme Treatment of crushed fruit mass is applied to some fruits by adding 2-8% pectolitic enzymes at about 50° C for 30 minutes.

This optional step has the following advantages: extraction yield will be improved, the juice colour is better fixed and finished product taste is improved.

However, for fruit which is naturally rich in pectic substances, this treatment makes the resulting "exhausted" material useless for industrial pectin production.

Heating of crushed fruit mass before juice extraction is an optional step used for some fruit in order to facilitate pressing and colour fixing; at same time, protein coagulation takes place.

Pressing to extract juice.

Diffusion is an alternative step for juice extraction and can be carried out discontinuously or in batteries at water temperature of about 80-85°C.

Juice Clarifying can be performed by centrifugation or by enzyme treatment. Centrifugation achieves a separation of particles in suspension in the juice and can be considered as a pre-clarifying step. This operation is carried out in centrifugal separators with a speed of 6000 to 6500 RPM.

Enzyme clarifying is based on pectic substance hydrolysis; this will decrease the juices' viscosity and facilitate their filtration. The treatment is the addition of pectolitic enzyme preparations in a quantity of 0.5 to 2 g/l and will last 2 to 6 hours at room temperature, or less than 2 hours at 50° C, a temperature that must not be exceeded.

The control of this operation is done by checking the decrease in juice viscosity. Sometimes, the enzyme clarifying is completed with the step called "sticking" by the addition of 5-8 g/hl of food grade gelatine which generates a flocculation of particles in suspension by the action of tannins.

Filtration of clarified juice can be carried out with kieselgur and bentonite as filtration additive in press-filters (equipment).

De-tartarisation is applied only to raisin juice and is aimed to eliminate potassium bitartrate from solution. This step can be performed by the addition of 1% calcium lactate or 0.4% calcium carbonate.

Pasteurization of juice can be done for temporary preservation (pre-pasteurization) and in this case this operation is carried out with continuous equipment (heat exchangers, etc.); warm juice is stored in drums or large size receptacles (20-30 kg). Pasteurization conditions are at 75°C in continuous stream.

Pasteurization of bottled juice is then carried out just before delivery to the market; this is performed in water baths at 75° C until the point where the juice reaches 68° C. In cases when the final pasteurization is done without pre-pasteurization and temporary storage, modern methods use a rapid pasteurization followed by aseptic filling in receptacles.

Rapid pasteurization conditions are as follows: temperature about 80° C, over 10-60 sec., followed by cooling; all operations are carried out in continuous stream.

Preservation under CO_2 pressure may be done at a concentration of 1.5% CO_2 under a pressure of 7 kg/cm^2. At the distribution step, proceed at CO_2 decompression and the juice is then submitted to a sterilising filtration and aseptic filling in receptacles.

Preservation by freezing is carried out at about -30° C, after a preliminary de-aeration; storage is at -15 to -20° C.

Production of concentrated juices by evaporation is performed under vacuum (less than 100 mm Hg residual pressure) up to a concentration of 65-70% total sugar which assures preservation without further pasteurization. Modern evaporation installations recover flavours from juices which are then reincorporated in concentrated juices.

Additional operations for juice manufacturing are the vacuum de-aeration and mixing with other fruit juices or with sugar.

For the production of non clarified juices the centrifugation is the only specific step, enzyme clarifying and subsequent filtration being eliminated.

The optimum sugar/acid ratio for the majority of fruit, mainly for pomaces, is 10/1 to 15/1.

Fruit which is rich in carotenoids (apricots, peaches, etc.) is only processed as juices with pulp ("nectars").

Technological flow-sheet for fruit juices with pulp ("Nectars")

This process is divided at industrial scale in two categories of operations:

a. processing for obtaining juices;
b. juice conditioning for preservation.

(a) Operations in the first category are differ according to the type of fruit which to be processed.

Pomaces (apples, pears) are washed and sorted and then crushed in a colloid mill; fruit puree is then passed through a screw type heating equipment where direct steam is used as a source of heat. Warm fruit mass is treated in a pulper with a 2 mm screen and then through an extractor similar with the equipment used for tomato juice.

Stone fruits (apricots, peaches, cherries, etc.) after washing and sorting are submitted to steam in a continuous heater, then the warm fruit mass is passed through a pulper and then an extractor (as mentioned above). Berries (strawberry, wild berries, etc.) are washed, sorted and then crushed, preheated and then introduced in extractor.

In order to avoid browning and undesirable taste modifications it is usual to add about 0.05% ascorbic acid.

(b) Second category type of operations are similar for all fruit species. Partial elimination of cellulose is achieved with a continuous centrifugal separator; the resulting juice is then processed in order to adjust sugar and acid content for viscosity.

Sugar (about 8-10%) is added as a syrup (in water or in the juice of same fruit obtained by pressure). Acidity is adjusted with citric or tartaric acid. The adjusted juice is then deaerated under vacuum at about 40° C. This step aims at avoiding oxidative reactions and vitamin C loss reduction.

An important subsequent step is an intensive homogenisation (under pressure at 150-180 A) in order to obtain particles with dimensions below 100. The homogenised juice obtained is then continuously pasteurized in plate heat exchanger equipment at a temperature of about 130° C, cooled down to about 90° C and aseptically packed in receptacles.

The principal characteristics of fruit "nectars" are uniformity and stability of the content provided by the advanced disintegration of fruits. Stability can be obtained by increasing product viscosity by adding pectin for fruit which is deficient in this component. In order to avoid "separation", intensive homogenisation is carried out as described above. Fruit "nectars" contain all the important components of the original fruit and to a large extent maintain their taste and flavour. The sugar/acidity (as citric acid) ratio is to a large extent determined by the type of fruit and the correction applied; for example, this ratio is 30 for apricots, 40 for peaches, 160 for pears, etc. A general technological flow-sheet for fruit juice processing is presented below.

Banana and Plantain Processing Technologies

Traditional Processing

1. *Products:* Uses and Dietary Significance : Most of the world's bananas are eaten either raw, in the ripe state, or as a cooked vegetable, and only a very small proportion are processed in order to obtain a storable product. This is true both at a traditional village level with both dessert and cooking bananas and when considering the international trade in dessert bananas.

   In general, preserved products do not contribute significantly to the diet; however, in some localised areas the products are important in periods when food are scarce.

   Probably the most widespread and important product is flour preparation from unripe banana and plantains by sun-drying. In Uganda, dried slices known as "mutere" are prepared for storage from green bananas, the dried slices being either used directly for cooking or after grinding into a flour. "Mutere" is used chiefly as a famine reserve and does not feature largely in the diet under normal conditions.

   In Gabon, plantains are sometimes made into dried slices which can be stored and used on long journeys, and plantains are used in Cameroon to prepare dried pieces which are stored and ground as needed into flour for use in cooking a paste known as "fufu". Dried green banana slices are also used in parts of South and Central America and West Indies for preparing flour.

   The other nutritionally important product is beer which is a major product in Uganda, Rwanda and Burundi where green banana utilisation is particularly high.

2. Preservation Methods and Processes.

   Drying. Both ripe and unripe bananas and plantains are normally peeled and sliced before drying, although banana figs are sometimes prepared from whole ripe fruit. Sun drying is the most widespread technique where the climate is suitable but drying in ovens or over fires is also practiced. In west Africa, plantains are often soaked and sometimes parboiled before drying.

The slices of unripe fruit are normally spread out on bamboo frameworks; or a cemented area; or on a mat; or on a swept-bare patch of earth; or on a roof; or sometimes on stones outcrops or sheets of corrugated iron. Oven-drying of ripe bananas is practiced in Polynesia as a mean of preserving the fruits, which are then wrapped in leaves and bound tightly to store until needed. In East Africa a method has been reported that involves drying the peeled bananas on a frame placed over a fire for 24 hr before drying in the sun, to accelerate the process.

3. Product stability and storage problems : There is little experimental data on the storage life of the traditionally made banana and plantain products.
4. Potential for scaling up of traditional processes to industrial level : Many banana products are now produced on an industrial scale, including the traditional banana figs and flour, and the processing techniques are described below. One of the main problems encountered has been the susceptibility of banana products to flavour deterioration and discoloration and in the past many products reaching the market have been of poor quality.

A great deal of research has been directed to overcoming these problems, although however good the resultant products are they cannot compare in flavour and other characteristics with the fresh banana fruit. Indeed, an important constraint on the large-scale development of banana processing is the lack of demand for banana products since the fresh fruit is available throughout the year in most parts of the tropical world.

The production of beer from banana and plantains has not been scaled up to an industrial level, and while an important product in localised areas of tropical Africa, the market is rapidly declining in favour of European-type brews produced locally.

Industrial Processing

Products and Uses: The main commercial products made from bananas are canned or frozen puree, dried figs, banana powder, flour, flakes, chips (crisps), canned slices and jams.

Banana products can be divided roughly into two types - those for direct consumption, such as figs, and those for use in food manufacturing industry, for example purees and powder.

Banana figs, or fingers as they are sometimes known, are usually whole, peeled fruit carefully dried so as to retain their shape, although sometimes the fruit is sliced or halved to facilitate drying. Banana and plantain chips (crisps) are thinly sliced pieces of fruit fried in oil and eaten as a snack like potato chips (crisps).

The main use of canned slices is in tropical fruit salads. Banana flakes are used as a flavouring or in breakfast cereals. Banana puree find use mainly in the production of baby foods. Banana flour is said to be highly digestible and is used in baby and invalid foods, but can also be used in the preparation of bread and beverages.

Banana powder is used chiefly in the baking industry for the preparation and fillings for cakes and biscuits and is also used for invalid and baby foods.

Processing Technology: In general, to obtain a good-quality product from ripe-bananas the fruit is harvested green and ripened artificially under controlled conditions at the processing factory. After ripening, the banana hands are washed to remove dirt and any spray residues, and peeled. Peeling is almost always done by hand using stainless steel knives, although a mechanical peeler for ripe bananas has been developed, capable of peeling 450 Kg of fruit per hour (Banana Bulletin, 1974).

The peeling of unripe bananas and plantains is facilitated by immersing the fruit in hot water. For example, immersion in water at 70-75 ° C for 5 min. has been suggested as an aid for peeling green bananas for flour production, while the peeling of green bananas for freezing has been facilitated by immersion in water at 93° C for 30 min.

Banana Figs: Fully ripe fruits with a sugar content of about 19.5% are used and are treated with sulphurous acid after peeling, then dried as soon as possible afterwards. Various drying systems have been described using temperatures between 50 and 82° C for 10 to 24 hr to give a moisture content ranging from 8 to 18% and a yield of dried figs of 12 to 17% of the fresh banana on the stem.

One factory in Australia uses a solar heat collector on the roof to augment the heat used for drying bananas. Bananas can also be dried by osmotic dehydration, using a technique which involves drawing water from 1/4-in. thick banana by placing them in a sugar solution of 67 to 70 deg. Brix for 8 to 10 hr. followed by vacuum-drying at 65 to 70° C, at a vacuum of 10 mm Hg for 5 hr. The moisture content of the final products is 2.5% or less, much lower than that achieved by other methods.

Banana Puree: Banana puree is obtained by pulping peeled, ripe bananas and then preserving the pulp by one of three methods: canning aseptically, acidification followed by normal canning, or quick-freezing.

The bulk of the world's puree is processed by the aseptic canning technique. Peeled, ripe fruits are conveyed to a pump which forces them through a plate with 1/4-in. holes, then onto a homogeniser, followed by a centrifugal de-aerator, and into a receiving tank with 29in. vacuum, where the removal of air helps prevent discoloration by oxidation.

The puree is then passed through a series of scraped surface heat exchangers where it is sterilised by steam, partially cooled, and finally brought to filling temperature. The sterilised puree is then packed aseptically into steam-sterilised cans which are closed in a steam atmosphere.

Banana Slices: Several methods for canning of banana slices in syrup are used. Best-quality slices are obtained from fruit at an early stage of ripeness. The slices are processed in a syrup of 25 deg. Brix with pH about 4.2, and in some processes calcium chloride (0.2%) or calcium lactate (0.5%) are added as firming agents. A method for producing an intermediate-moisture banana product for sale in flexible laminate pouches has been developed. Banana slices are blanched and equilibrated in a solution containing glycerol (42.5%), sucrose (14.85%), potassium sorbate (0.45%), and potassium metabisulphite (0.2%) at 90 deg. C for 3 min. to give a moisture content of 30.2%.

Banana Powder: In the manufacture of banana powder, fully ripe banana pulp is converted into a paste by passing through a chopper followed by a colloid mill. A 1 or 2 % sodium metabisulphite solution is added to improve the colour of the final product. Spray- or drum-drying may be used, the latter being favoured as all the solids are recovered.

A typical spray dryer can produce 70 kg powder per hour to give yields of 8 to 11% of the fresh fruit, while drum-drying gives a final yield of about 13% of the fresh fruit. In the latter method the moisture content is reduced to 8 to 12 % and then further decreased to 2 % by drying in a tunnel or cabinet dryer at 60° C.

Banana Flour: Production of flour has been carried out by peeling and slicing green fruit, exposure to sulphur dioxide gas, then drying in a counter-current tunnel dryer for 7 to 8 hr. with an inlet temperature of 75° C and outlet temperature of 45° C, to a moisture content of 8%, and finally milling.

Banana Chips (crisps): Typically, unripe peeled bananas are thinly sliced, immersed in a sodium or potassium metabisulphite solution, fried in hydrogenated oil at 180 to 200° C, and dusted with salt and an antioxidant.

Alternatively, slices may be dried before frying and the antioxidant and salt added with the oil. Similar processes for producing plantain chips have been developed.

Banana Beverages: In a typical process, peeled ripe fruit is cut into pieces, blanched for 2 min. in steam, pulped and pectolytic enzyme added at a concentration of 2 g enzyme per 1 kg pulp, then held at 60 to 65° C and 2.7 to 5.5 pH for 30 min.

In a simpler method, lime is used to eliminate the pectin. Calcium oxide (0.5%) is added to the pulp and after standing for 15 min. this is neutralised giving a yield of up to 88% of a clear, attractive juice. In another process banana pulp is acidified, and steam-blanched in a 28-in Hg vacuum which ensures disintegration and enzyme inactivation. The pulp is then conveyed to a screw press, the resulting puree diluted in the ratio 1:3 with water, and the pH adjusted by further addition of citric acid to 4.2 to 4.3, which yields an attractive drink when this is centrifuged and sweetened.

Jam: A small amount of jam is made commercially by boiling equal quantities of fruit and sugar together with water and lemon juice, lime juice or citric acid, until setting point is reached.

Product Stability and Spoilage Problems

All dried banana products are very hydroscopic and susceptible to flavour deterioration and discoloration, but this can be overcome to some extent by storing in moisture-proof containers and sulphiting the fruit before drying to inactivate the oxidases. The dried products are also liable to attack by insects and moulds if not stored in dry conditions, although disinfestation after drying by heating for 1 hr to 80° C or by fumigation with methyl bromide ensures protection against attack. Banana powder is said to be stored for up to a year commercially and flakes have been stored in vacuum-sealed cans with no deterioration in moisture, colour or flavour for 12 months. Banana chips tend to have a poor storage life and to become soft and rancid. However, chips treated with an antioxidant have been stored satisfactorily at room temperature in hermetically sealed containers up to 6 months with no development of rancidity.

Quality Control Methods

In general a good quality product is obtained if fruit is harvested at the correct stage of maturity and, where appropriate, ripened under controlled conditions. For example, in the case of banana figs, the fruit should be fully mature (sugar content of 19.5% or above) or the final product is liable to be tough and lacking in flavour. However, if over-ripe fruit is used, the figs tend to be sticky and dark in colour, so the fruit must be fully yellow but still firm.

For banana flour, which is prepared from unripe bananas, the fruit is harvested at three-quarters the full-ripe stage and is processed within 24 hr. prior to the onset of ripening. If less mature fruit is used, the flour tastes slightly astringent and bitter due to the tannin content. Bananas harvested between 85 and 95 days after the emergence of the inflorescence, with a pulp-to-peel ratio of about 1.7, were considered to be most suitable for the deep-fat frying.

Other criteria suggested for assessing maturity were beta-carotene and reducing sugar content, both of which increase with increasing maturity and pH which decreases as the fruit ripens, and these should be, respectively, about 2000 μg/100 g, less than 1.5% and 5.8 or above. Browning was found to occur if the sugar content was higher than 1.5%. The determination of crude fat in processed chips is also considered to be a necessary quality control measure. It is important to remove all impurities prior to processing of products, and this is done by washing to remove dirt and spray residues and control on the processing line so that substandard fruit can be removed.

Preparation Methods for Fresh Bananas and Plantains

The main ways of preparing fresh bananas for consumption are boiling or steaming, roasting or baking and frying. Boiling followed by pounding into "fufu" is also widely adopted in certain areas of the tropics.

Boiling or Steaming: Plantains and bananas are often prepared simply by boiling in water, either in their peel or after peeling, and either ripe or unripe; if unripe, the fruit is scraped thoroughly after peeling to remove all traces of fibrous material. The boiled fruit is eaten alone or more usually accompanied by a sauce. This preparation technique is widely used in West Africa.

Roasting or Baking: Unpeeled or peeled fruit, either ripe or unripe, is roasted simply by placing in the ashes of a fire or in an oven. This method is widely used in West Africa, East Africa and the South Pacific islands. For example, ripe plantains are placed unpeeled in an oven and when partly brown and tender, removed and peeled, then replaced in the oven and roasted evenly.

Frying: Ripe or unripe plantains or bananas are often peeled, sliced and cooked in oil, particularly in West Africa and in parts of South America and the West Indies. Similar products are also made in East Africa. Typically, ripe plantains are peeled, cut into slices or split lengthways, and fried in palm oil or with groundnut oil, the pieces being served either hot with a sauce or with fried eggs, or cold as a snack.

Pounding: Pounding is a process, used particularly in West Africa, for preparing most perishable staple food crops including plantains, cassava, yams and cocoyams to obtain a paste or dough known as "fufu" (also spelled "foofoo", "foutou", "foufou"). The plantains are peeled or boiled and peeled after boiling and pounded in a wooden mortar, the resulting paste normally being eaten with soup or a spiced sauce of meat and vegetables, but sometimes after wrapping in leaves and steaming.

Mango and Guava Processing Technologies

Mangoes are processed at two stages of maturity. Green fruit is used to make chutney, pickles, curries and dehydrated products. The green fruit should be freshly picked from the tree. Fruit that is bruised, damaged, or that has prematurely fallen to the ground should not be used. Ripe mangoes are processed as canned and frozen slices, puree, juices, nectar and various dried products. Mangoes are processed into many other products for home use and by cottage industry.

The mango processing presents many problems as far as industrialization and market expansion is concerned. The trees are alternate bearing and the fruit has a short storage life; these factors make it difficult to process the crop in a continuous and regular way. The large number of varieties with their various attributes and deficiencies affects the quality and uniformity of processed products.

The lack of simple, reliable methods for determining the stage of maturity of varieties for processing also affects the quality of the finished products. Many of the processed products require peeled or peeled and sliced fruit. The lack of mechanised equipment for the peeling of ripe mangoes is a serious bottleneck for increasing the production of these products.

A significant problem in developing mechanised equipment is the large number of varieties available and their different sizes and shapes. The cost of processed mango products is also too expensive for the general population in the areas where most mangoes are grown. There is, however, a considerable export potential to developed countries but in these countries the processed mango products must compete with established processed fruits of high quality and relatively low cost.

Green Mango Processing

Pickles: The optimum stage of maturity should be determined for each variety used to make pickles.

There are two classifications of pickles - salt pickles and oil pickles. They are processed from whole and sliced fruit with and without stones. Salt is used in most pickles.

The many kinds of pickles vary mainly in the proportions and kinds of spices used in their preparation. One basic recipe for the study of the preparation and storage of pickles in oil is as follows:

Mango pieces	250 g	Tumeric powder	2 to 4 g
Salt	60 g	Fenugreek seeds	2 to 4 g
Mustard powder	30 g	Bengal gram seeds	2 to 4 g
Chili powder	20 g	Gingelly oil	20 to 30 g

The ingredients are mixed together and filled into wide-mouthed bottles of 0.5 kg capacity. Three days later the contents are thoroughly mixed and refilled into the bottles. Extra oil is added to form a 1-2 cm layer over the pickles.

Chutney: The product is prepared from peeled, sliced or grated unripe or semi-ripe fruit by cooking the shredded fruit with salt over medium heat for 5 to 7 minutes, mixed and then sugar, spices and vinegar are added. Cook over moderate heat until the product resembles a thick puree, add remaining ingredients and simmer another 5 min. Cool and preserve in sterilised jars. Spices usually include cumin seeds, ground cloves, cinnamon, chili powder, ginger and nutmeg. Other ingredients such as dried fruits, onions, garlic and nuts may be added.

Drying/dehydration: Immature fruit is peeled and sliced for sun-drying. The dried mango slices can be powdered to make a product called amchoo. The use of blanching, sulphuring and mechanical dehydration gives a product with better colour, nutrition, storability and fewer microbiological problems.

Ripe Mango Processing

Puree: Mangoes are processed into puree for re-manufacturing into products such as nectar, juice, squash, jam, jelly and dehydrated products. The puree can be preserved by chemical means, or frozen, or canned and stored in barrels. This allows a supply of raw materials during the remainder of the year when fresh mangoes are not available.

It also provides a more economical means of storage compared with the cost of storing the finished products, except for those which are dehydrated, and provides for more orderly processing during peak availability of fresh mangoes.

Mangoes can be processed into puree from whole or peeled fruit. Because of the time and cost of peeling, this step is best avoided but with some varieties it may be necessary to avoid off-flavours which may be present in the skin. The most common way of removing the skin is hand-peeling with knives but this is time-consuming and expensive. Steam and lyepeeling have been accomplished for some varieties.

Several methods have been devised to remove the pulp from the fresh ripe mangoes without hand-peeling. A simplified method is as follows: the whole mangoes were exposed to atmospheric steam for 2 to 2 1/2 min in a loosely covered chamber, then transferred to a stainless steel tank.

The steam-softened skins allowed the fruit to be pulped by a power stirrer fitted with a saw-toothed propeller blade mounted 12.7 to 15.2 cm below a regular propeller blade. The pulp is removed from the seeds by a continuous centrifuge designed for use in passion fruit extraction. The pulp material is then passed through a paddle pulper fitted with a 0.084 cm screen to remove fibre and small pieces of pulp.

Mango puree can be frozen, canned or stored in barrels for later processing. In all these cases, heating is necessary to preserve the quality of the mango puree. In one process, puree is pumped through a plate heat exchanger and heated to 90°C for 1 min and cooled to 35° C before being filled into 30 lb tins with polyethylene liners and frozen at -23.50 C.

In an other process, pulp is acidified to pH 3.5, pasteurized at 90°C, and hot-filled into 6 kg high-density bulk polyethylene containers that have been previously sterilised with boiling water. The containers are then sealed and cooled in water. This makes it possible to avoid the high cost of cans.

Wooden barrels may be used to store mango pulp in the manufacture of jams and squashes. The pulp is acidified with 0.5 to 1.0% citric acid, heated to boiling, cooled, and SO_2 is added at a level of 1000 to 1500 ppm in the pulp. The pulp is then filled into barrels for future use.

Slices: Mango slices can be preserved by canning or freezing, and recent studies have shown the feasibility of pasteurized-refrigerated and dehydro-canned slices. The quality of the processed product in all of these procedures will be dependent upon selection of a suitable variety along with good processing procedures. Thermal process canning of mango slices in syrup is the most widely used preservation method.

Beverages: The commercial beverages are juice, nectar and squash. Mango nectar and juice contain mango puree, sugar, water and citric acid in various proportions depending on local taste, government standards of identity, pH control, and fruit composition of the variety used. Mango squash in addition to the above may contain SO2 or sodium benzoate as a preservative. Other food grade additives such as ascorbic acid, food colouring, or thickeners may be used in mango beverages.A short description of finished products found in literature is as follows:

- *Mango Juice:* prepared by mixing equal quantities of pulp (puree) and water together and adjusting the total soluble solids (TSS) and acidity to taste (12 to 15% TSS and 0.4 to 0.5% acidity as citric acid);
- Mango nectar containing 25% puree can be prepared using the following procedure.

	Brix of purée		
Nectar components	15°	17°	20°
Purée	100	100	100
Sugar	45	43	40
Water	255	257	260

Commercial processing conditions may require the use of a preservative.

The pH is adjusted to approximately 3.5 by adding citric acid as a 50% solution.

The time of heat processing will vary with filling temperature, can size and viscosity of the juice or nectar.

Mango squash may be prepared according to flow-sheet described below; the finished product may contain 25% juice, 45% TSS and 1.2 to 1.5% acidity and may be preserved with sulphur dioxide (350ppm) or sodium benzoate (1000 ppm) in glass bottles.

Mango squash simplified flow-sheet.

Ingredients

Mango pulp	900	900
Sugar	900	1100
Citric acid	18	15
Water	900	900

Mangoes are washed, stored, peeled with stainless steel knives. The pulp is prepared by using a pulper with fine sieve (0.025-in); Sugar is mixed with water and citric acid = syrup; The pulp is added to the syrup and mixed well; The mixture is strained trough cloth; The squash is heated at 85° C and bottles are filled and closed. For additional heat treatment bottles may need to be maintained at a product temperature of 80°C for 30 minutes if the product is to be processed without preservatives. The bottles are then left to cool in water and stored at room temperature.

Two negative points must be avoided: presence of air bubbles (which is a source of quick deterioration) and separation of squash solids (giving an undesirable appearance). The means to avoid these two phenomena are described in the fruit juices section. A type of "squash type" beverage may also be manufactured with '/ a pulp + '/a water + i/a sugar and pH adjusted to 3.7 by addition of citric acid. Using different sieve sizes affects the quality and reduces air bubbles to a certain extent but homogenisation and de-aeration of puree or squash seem to be important in order to avoid separation and air bubbles.

The squash quality is evaluated on the basis of the following characteristics: pH, titrable acidity, soluble solids, ascorbic acid (by 2,6 dichlorophenol indophenol method), specific gravity.

Dried/dehydrated: Ripe mangoes are dried in the form of pieces, powders, and flakes. Drying procedures such as sun-drying, tunnel dehydration, vacuum-drying, osmotic dehydration may be used. Packaged and stored properly, dried mango products are stable and nutritious.

One described process involves as pre-treatment dipping mango slices for 18 hr (ratio 1:1) in a solution containing 40° Brix sugar, 3000 ppm SO_2, 0.2% ascorbic acid and 1% citric acid; this method is described as producing the best dehydrated product. Drying is described using an electric cabinet through flow dryer operated at 60° C. The product showed no browning after 1 year of storage.

Drum-drying of mango puree is described as an efficient, economical process for producing dried mango powder and flakes. Its major drawback is that the severity of heat preprocessing can produce undesirable cooked flavours and aromas in the dried product. The drum-dried products are also extremely hydroscopic and the use of in-package desiccant is recommended during storage.

Canning: This preservation technology is described in various technological flow-sheets in this bulletin. Mango bar or "fruit leather" is presented in various flow-sheets.

Guava Processing Technologies

Guava Puree: Guava puree is used in the manufacture of guava nectar, various juice drink blends and in the preparation of guava jam. The washed sound fruit is first passed through a chopper or slicer to break up the fruit and this material is fed into a pulper. The pulper will remove the seeds and fibrous pieces of tissue and force the reminder of the product through a perforated stainless steel screen. The holes in the screen should be between 0.033 and 0.045 in. The machine should be fed at a constant rate to ensure efficient operation. The pureed material coming from the pulper is next passed through a finisher. The finisher is equipped with a screen containing holes of approximately 0.020 in. The finisher will remove the stone cells from the fruit and provide the optimum consistency to the product. Perhaps the best way to preserve the guava puree is by freezing and the material passing through the finisher can be packaged and frozen with no further treatment. It is not necessary to heat the product to inactivate enzymes or for other purposes. The material can be frozen in a number of types of cartons and cans; however, a fibre box with a plastic bag inside is commonly used and is probably the less expensive.

It is also possible to can and heat process the guava puree and this can be accomplished by heating the puree to 195° F in an open double bottom kettle, filling into cans, closing the cans, inverting the cans for a few seconds, followed by cooling. Cans should be cooled rapidly to approximately 100° F before they are cased and stacked into warehouses.

Guava Juice and Concentrate: Guava juice can be used in the manufacture of a clear guava jelly or in various drinks. A clear juice may be prepared from guava puree that is depectinised enzymatically. About 0.1% pectin-degrading enzyme is mixed into the puree at room temperature; heating of the product at approximately 120° F will greatly speed the action of the enzyme. After 1 hr. clear juice is separated from the red pulp by centrifuging or by pressing in a hydraulic juice press. A batch-type or continuous-flow centrifuge can be used on the depectinised puree with no further treatment.

The clear juice after centrifuge or after press (and subsequent filtration) can be preserved by freezing or by pasteurization in hermetically sealed cans. For shipment to overseas markets it may be advantageous to concentrate either the puree or the juice.

Recent Trends in Fruit and Vegetable Processing

New Products: The number and variety of fruit and vegetable products available to the consumer has increased substantially in recent years. The fruit and vegetable industry has undoubtedly benefited from the increased recognition and emphasis on the importance of these products in a healthy diet. Traditional processing and preservation technologies such as heating, freezing and drying together with the more recent commercial introduction of chilling continue to provide the consumer with increased choice. This has been achieved by new heating (*e.g.* UHT, microwave, ohmic) and freezing (*e.g.* cryogenic) techniques combined with new packaging materials and technologies (*e.g.* aseptic, modified atmosphere packaging).

The overall trend in new fruit and vegetable products is "added value", thus providing increased convenience to the consumer by having much greater variety of ready prepared fruit and vegetable products. These may comprise complete meals or individual components. The suitability of products and packages for microwave re-heating has been an important factor with respect to added convenience. The major trends in the development of fruit and vegetable heat processed products in recent years are shown in table; the number of new fruit and vegetable products.

Numbers of New Fruit and Vegetable Products

	1990	1991	1992 Jan-June
Frozen vegetable products	66	95	21
Chilled vegetable products	76	81	78
Heat processed vegetables	51	60	38
Heat processed fruits	13	14	5
Fruit juices and drinks	73	83	46
Potato crisps	32	33	16

Source: C. Dennis (1993)

New product development in the fruit and vegetable sector is most important in meeting the continued challenge of providing the consumer with choice and high quality products.

A Fresh Look at Dried Fruit

New fruit varieties and advance in drying technologies are putting a fresh twist on dried fruit applications. Fruits that have been introduced to the drying process include cranberries, blueberries, cherries, apples, raspberries and strawberries - not to mention the traditional mainstays of raisins, dates, apricots, peaches, prunes and figs.

Perceived as a "value-added" ingredient, dried fruit adds flavour, colour, texture and diversity with little alteration to an existing formula. The growing interest in ethnic cuisines in U.S.A. and the change to a more healthy way of eating, has also moved dried fruit considerably closer to the mainstream.

Found primarily in the baking industry, dried fruit is coming into its own in various food products, including entrees, side dishes and condiments. Compotes, chutneys, rice and grain dishes, stuffings, sauces, breads, muffins, cookies, deserts, cereals and snacks are all food categories encompassing dried fruit.

Since some dried fruit is sugar infused (osmotic drying), food processors can decrease the amount of sugar in formula- this is especially the case in baked products. Processors are making adjustments in moisture content of the dried fruit so that a varied range is available for different applications. An added bonus is dried fruits' shelf stability (a shelf life of at least 12 months). Dried fruit is more widely available in different forms, including whole dried, cut, diced and powders.

Citric acid and its use in fruit and vegetable processing.

Citric acid may be considered as "Nature's acidulant".

It is found in the tissues of almost all plants and animals, as well as many yeasts and moulds.

Commercially citric acid is manufactured under controlled fermentation conditions that produce citric acid as a metabolic intermediate from naturally-occurring yeasts, moulds and nutrients. The recovery process of citric acid is through crystallization from aqueous solutions.

Citric acid is widely used in carbonated and still beverages, to impart a fresh-fruit "tanginess". Citric acid provides uniform acidity, and its light fruity character blends well and enhances fruit juices, resulting in improved palatability. The amount of citric acid used depends on the particular desired flavour (*e.g.*, High-acid: lemonade; Medium-acid: orange, punch, cherry; Low-acid: strawberry, black cherry, grape).

Sodium citrate is often added to beverages to mellow the tart taste of high acid concentrations. It provides a cool, distinctive smooth taste and masks any bitter aftertaste of artificial sweeteners. In addition, it serves as a buffer to stabilise the pH at the desired level. The high water solubility of citric acid (181 g/100 ml) makes it an ideal additive for fountain fruit syrups and beverages concentrates as a flavour enhancer and microbial growth inhibitor (preferably at pH < 4.6).

In processed fruits and vegetables, citric acid performs the following functions:

a. It reduces heat-processing requirements by lowering pH: inhibition of microbial growth is a function of pH and heat treatment. Higher heat exposure and lower pH result in greater inhibition. Thus the use of citric acid to bring pH below 4.6 can reduce the heating requirements. In canned vegetables, citric acid usage is greatest in tomatoes, onions and pimentos. For tomato packs, the National Canners Association recommends a pH of 4.1 to 4.3. In general, 0. 1% citric acid will reduce the pH of canned tomatoes by 0.2 pH units.
b. Optimise flavour: citric acid is added to canned fruits to provide for adequate tartness. Recommended usage level is generally less then 0.15%.
c. Supplement antioxidant potential: citric acid is used in conjunction with antioxidants such as ascorbic and erythorbic acids, to inhibit colour and flavour deterioration caused by metal-catalysed enzymatic oxidation. Recommended usage levels are generally 0.1% to 0.3% with the antioxidant at 100 to 200 ppm.
d. Inactivate undesirable enzymes: oxidative browning in most fruits and vegetables is catalysed by the naturally present polyphenol oxidase. The enzymatic activity is strongly dependent on pH.

Addition of citric acid to reduce pH below 3 will result in inactivation of this enzyme and prevention of browning reactions.

Cherry and Apricot Oils are Safe for Food Use

The oils obtained by cold pressing the kernels of the cherry (Prunus cerasus) and the apricot (Prunus armeniaca) have been declared acceptable for food use by the UK Ministry of Agriculture, Fisheries & Food's Advisory Committee on Novel Foods and Processes (ACNFP) by June 1993.

In its assessments of the safety in use of the cherry and apricot kernel oils, the ACNFP consider specifications that included data on fatty acid composition, the presence of natural antioxidants and the content of cyanide, mycotoxins and heavy metals.

The Committee says that it gave particular consideration to the possible presence in the oils of the cyanogenic glucoside amygdalin, from which cyanide is released by enzymic action when the kernels of cherry and apricot are crushed. Amygdalin was found to be absent from the cherry and apricot kernel oils.

The ACNFP is satisfied that there are no food safety reasons why the use of cherry and apricot kernel should not be acceptable provided there is compliance with the specifications.

Specification of Purity for Cherry and Apricot Kernel Oils as Determined by UK ACNFP

	Cherry	Apricot
Contaminants limits:		
Heavy metals (total)	0.5 mg/kg	0.5 mg/kg
Aflatoxins (total)	4.0 g/kg	4.0 g/kg
Cyanide	0.15 mg/kg	0.15 mg/kg
Pesticide residues	0.01 mg/kg	0.01 mg/kg
Tocopherols:		
alpha/delta/gamma (mg/kg)	356-886	569-899

Source: Anon. (1993)

The oils are obtained by the mechanical mincing and cold pressing of kernels extracted from cleaned cherry or apricot stones. After filtering, the oils are stored and are to be sold in a raw, unrefined state. The cherry and apricot kernel oils are high unsaturated and are expected to be used as speciality oils for salad dressings, baking and shallow frying applications.

The Use of Fruit Juices in Confectionery Products

During the last decade, the concept of fruit juices has gained immensely on consumer popularity. The majority of new non-alcoholic and alcoholic fruit drink products were a combination of syrups, fruit juices and flavours. The confectionery industry followed suit and new products incorporated fruit juices as part of their confectionery formulations and processes. Fruit juice concentrates of high solids are often used instead of normal or single-fold juices.

Juice concentrates are made of pure fruit juices. The process starts with pressing fruits and obtaining pure fruit juice; this is stabilised by heat treatment which inactivates enzymes and microorganisms. The next processing step is concentration under vacuum up to 40-65° Brix or 4-7 fold. The concentrates are then blended for standardisation and stored.

These fruit juice concentrates are often further stabilised by the addition of sodium benzoate and potassium sorbate and are usually stored away from light and are refrigerated or frozen.

Depectinised fruit juices are also used to prevent foaming in confectionery processes and are essential for use in clear beverage products. Fruit juice concentrates which are depectinised, and have added preservatives are called stabilised, clarified, fruit juice concentrates.

Fruit juices are used in confectionery products in conjunction with natural and artificial flavours which provides intense flavour impact and are cost-effective for a confectionery product.

The traditional concern in using fruit juice concentrates in confectionery applications has been the effect of the natural acids on the finished product, particularly the formation of invert sugar during processing.

This is a logical concern since concentrates contain differing amounts and types of acids. For example: apple, cherry, strawberry and other berries contain primarily malic acid. Grapes mainly contain tartaric acid. Cranberry is high in quinic acid. Citrus fruits and pineapple contain differing amounts of citric acid. The concentrates, when used, are normally buffered to a pH of 5-7 with sodium hydroxide.

In formulating products with fruit juice concentrates, the solids of the concentrate are considered as mostly reducing sugars and a reduction in corn syrup is made to compensate for equivalent amount of reducing sugar being added in the concentrate.

The exact replacement can be determined by measuring the D.E. of the concentrate to be added. In formulations when small amounts of concentrate are used (less than 1%), no adjustment is made since the reducing sugar contribution of the concentrate is not significant. Fruit juice concentrates can also be used to provide a source of natural colour, in particular red colour. Grape, raspberry, cherry, strawberry and cranberry concentrates in small amounts are very effective in colouring cream centres.

The inclusion of fruit juices in confectionery products is now left up to the imagination of the manufacturer. These products must, of course, hold up to the standards of flavour integrity, and product excellence, during the shelf-life of these products.

Vegetable Specific Processing Technologies

Vegetables Varieties

Vegetable processors must appreciate the substantial differences that varieties of a given vegetable will possess. In addition to variety and genetic strain differences with respect to weather, insect and disease resistance, varieties of a given vegetable will differ in size, shape, time of maturity, and resistance to physical damage.

Varietal differences then further extend into warehouse storage stability, and suitability for such processing methods as canning, freezing, pickling or drying. A variety of peas that is suitable for canning may be quite unsatisfactory for freezing and varieties of potatoes that are preferred for freezing may be less satisfactory for drying or potato chip manufacture. This should be expected since different varieties of a given vegetable will vary somewhat in chemical composition, cellular structure and biological activity of their enzyme system.

Harvesting and Pre-processing

When vegetables are maturing in the field they are changing from day to day. There is a time when the vegetable will be at peak quality from the stand-point of colour, texture and flavour.

This peak quality is quick in passing and may last only a day. Harvesting and processing of several vegetables, including tomatoes, corn and peas are rigidly scheduled to capture this peak quality. After the vegetable is harvested it may quickly pass beyond the peak quality condition. This is independent of microbiological spoilage; these main deteriorations are related to:

(a) loss of sugars due to their consumption during respiration or their conversion to starch; losses are slower under refrigeration but there is still a great change in vegetable sweetness and freshness of flavour within 2 or 3 days;

(b) production of heat when large stockpiles of vegetables are transported or held prior to processing.

At room temperature some vegetables will liberate heat at a rate of 127,000 kJ/ton/day; this is enough for each ton of vegetables to melt 363 kg of ice per day. Since the heat further deteriorates the vegetables and speeds microorganisms growth, the harvested vegetables must be cooled if not processed immediately.

But cooling only slows down the rate of deterioration, it does not prevent it, and vegetables differ in their resistance to cold storage. Each vegetable has its optimum cold storage temperature which may be between about 0-100 C (32-500 F).

(c) the continual loss of water by harvested vegetables due to transpiration, respiration and physical drying of cut surfaces results in wilting of leafy vegetables, loss of plumpness of fleshy vegetables and loss of weight of both.

Moisture loss cannot be completely and effectively prevented by hermetic packaging. This was tried with plastic bags for fresh vegetables in supermarkets but the bags became moisture fogged, and deterioration of certain vegetables was accelerated because of buildup of CO_2 and decrease of oxygen in the package.

It therefore is common to perforate such bags to prevent these defects as well as to minimise high humidity in the package which would encourage microbial growth.

Shippers of fresh vegetables and vegetable processors, whether they can, freeze, dehydrate, or manufacture soups or ketchup, appreciate the instability and perishability of vegetables and so do everything they can to minimise delays in processing of the fresh product. In many processing plants it is common practice to process vegetables immediately as they are received from the field.

To ensure a steady supply of top quality produce during the harvesting period the large food processors will employ trained field men; they will advise on growing practices and on spacing of plantings so that vegetables will mature and can be harvested in rhythm with the processing plant capabilities. This minimises stockpiling and need for storage.

Cooling of vegetables in the field is common practice in some areas. Liquid nitrogen-cooled trucks may next provide transportation of fresh produce to the processing plant or directly to market. Upon arrival of vegetables at the processing centre the usual operations of cleaning, grading, peeling, cutting and the like are performed using a moderate amount of equipment but a good deal of hand labour also still remains.

Reception: This covers qualitative and quantitative control of delivered vegetables. The organoleptic control and the evaluation of the sanitary state, even if they are very important steps in vegetables' characteristics assessment, cannot establish their technological value.

On the other hand, laboratory controls do not precisely establish their technological properties because of the difficulty in putting into showing some deterioration when using rapid control methods.

One correct method of vegetable quality appraisal is their overall evaluation based on the whole complex of data that can be obtained by combining an extensive organoleptic evaluation with simple analysis that can be performed rapidly in plant laboratory. These analysis can be:

a. refractometric extract (tomatoes, fruit, etc.);

b. specific weight (potatoes, peas, etc.);

c. consistency (measured with tenderometers, penetrometers, etc.);

d. boiling tests, etc.

Temporary storage.

This step should be as short as possible and better completely eliminated. Vegetables can be stored in:

a. simple stores, without artificial cooling;

b. in refrigerated stores; or, in some cases,

c. in silos (potatoes, etc.).

Simple stores should be covered, fairly cool, dry and well ventilated but without forced air circulation which can induce significant losses in weight through intensive water evaporation; air relative humidity should be at about 70-80%.

Refrigerated storage is always preferable and in all cases a processing centre needs a cold room for this purpose, adapted in volume I capacity to the types and quantities of vegetables (and fruits) that are further processed.

Washing: Washing is used not only to remove field soil and surface microorganisms but also to remove fungicides, insecticides and other pesticides, since there are laws specifying maximum levels of these materials that may be retained on the vegetable; and in most cases the allowable residual level is virtually zero. Washing water contains detergents or other sanitisers that can essentially completely remove these residues.

The washing equipment, like all equipment subsequently used, will depend upon the size, shape and fragility of the particular kind of vegetable:

- flotation cleaner for peas and other small vegetables;
- rotary washer in which vegetables are tumbled while they are sprayed with jets of water; this type of washer should not be used to clean fragile vegetables;

Sorting: This step covers two separate operations:

(a) Removal of non-standard vegetables (and fruit) and possible foreign bodies remaining after washing;

(b) Quality grading based on variety, dimensional, organoleptical and maturity stage criterion.

Skin Removal/peeling

Some vegetables require skin removal. This can be done in various ways.

(a) Mechanical: This type of operation is performed with various types of equipment which depend upon the result expected and the characteristics of the fruit and vegetables, for example:

i. a machine with abrasion device (potatoes, root vegetables);

ii. equipment with knives (apples, pears, potatoes, etc.);

iii. equipment with rotating sieve drums (root vegetables). Sometimes this operation is simultaneous with washing (potatoes, carrots, etc.) or preceded by blanching (carrots).

(b) Chemical: Skins can be softened from the underlying tissues by submerging vegetables in hot alkali solution. Lye may be used at a concentration of about 0.5-3%, at about 93° C (2000 F) for a short time period (0.5-3 min). The vegetables with loosened skins are then conveyed under high velocity jets of water which wash away the skins and residual lye.

In order to avoid enzymatic browning, this chemical peeling is followed by a short boiling in water or an immersion in diluted citric acid solutions.

It is more difficult to peel potatoes with this method because it is necessary to dissolve the cutin and this requires more concentrated lye solutions, up to 10%.

(c) Thermal:

Wet Heat (steam): Other vegetables with thick skins such as beets, potatoes, carrots and sweet potatoes may be peeled with steam under pressure (about 10 at) as they pass through cylindrical vessels. This softens the skin and the underlying tissue. When the pressure is suddenly released, steam under the skin expands and causes the skin to puff and crack. The skins are then washed away with jets of water at high pressure (up to 12 at).

Dry Heat (flame): Other vegetables such as onions and peppers are best skinned by exposing them to direct flame (about 1 min at 1000° C) or to hot gases in rotary tube flame peelers. Here too, heat causes steam to develop under skins and puff them so that they can be washed away with water.

Manual peeling only use when the other methods are impossible or sometimes as a completion of the other three ways. Average losses at this step are given in Table.

Losses at Vegetable Peeling, in %

	Peeling methods		
Vegetables	Manual	Mechanical	Chemical
Potatoes	15-19	18-28	-
Carrots	13-15	16-18	8-10
Beets	1416	13-15	9-10

Size Reduction: This step is applied according to specific vegetable and processing technology requirements.

Blanching: The special heat treatment to inactivate enzymes is known as blanching. Blanching is not indiscriminate heating. Too little is ineffective, and too much damages the vegetables by excessive cooking, especially where the fresh character of the vegetable is subsequently to be preserved by processing. This heat treatment is applied according to and depends upon the specificity of vegetables, the objectives that are followed and the subsequent processing / preservation methods.

Two of the more heat resistant enzymes important in vegetables are catalase and peroxidase. If these are destroyed then the other significant enzymes in vegetables also will have been inactivated.

The heat treatment to destroy catalase and peroxidase in different vegetables are known, and sensitive chemical tests have been developed to detect the amounts of these enzymes that might survive a blanching treatment. Catalase and peroxidase inactivation tests are presented in section 9.2.9.

Because various types of vegetables differ in size, shape, heat conductivity, and the natural levels of their enzymes, blanching treatments have to be established on an experimental basis. As with sterilisation of foods in cans, the larger the food item the longer it takes for heat to reach the centre. Small vegetables may be adequately blanched in boiling water in a minute or two, large vegetables may require several minutes.

Blanching as a unit operation is a short time heating in water at temperatures of 100° C or below. Water blanching may be performed in double bottom kettles, in special baths with conveyor belts or in modern continuous blanching equipment.

In order to reduce losses of hydrosoluble substances (mineral salts, vitamins, sugars, etc.) occurring during water blanching, several methods have been developed:

- temperature setting at 85-95° C instead of 100° C;
- blanching time has to be just sufficient to inactivate enzymes catalase and peroxidase;
- assure elimination of air from tissues.

An illustration of blanching parameters is seen in Table.

Blanching Parameters for Some Vegetables

Vegetables	*Temperature, °C*	*Time, min.*
Peas	85-90	2-7
Green beans	90-95	2-5
Cauliflower	Boiling	2
Carrots	90	3-5
Peppers	90	3

Steam heat treatment can also be applied instead of water blanching as a preliminary step before freezing or drying, as long as the preservation method is only used for enzyme inactivation and not to modify consistency.

For drying, the vegetables are conveyed directly from steaming equipment to drying installations without cooling. Vegetable steaming is carried out in continuous installations with conveyer belts made from metallic sieves. Cooling of vegetables after water blanching or steaming is performed in order to avoid excessive softening of the tissues and has to follow immediately after these operations; one exception is the case of vegetables for drying which can be transferred directly to drying equipment without cooling.

Natural cooling is not recommended because is too long and generates significant losses in vitamin C content. Cooling in pre-cooled air (from special installations) is sometimes used for vegetables that will be frozen.

Cooling in water can be achieved by sprays or by immersion; in any case the vegetables have to reach a temperature value under 37° C as soon as possible. Too long a cooling time generates supplementary losses in valuable hydrosoluble substances; in order to avoid this, the temperature of the cooling water has to be as low as possible.

Canning: Large quantities of vegetable products are canned. A typical flow sheet for a vegetable canning operation (which also applies to fruit for the most part) covers some food process unit operations performed in sequence: harvesting; receiving; washing; grading; heat blanching; peeling and coring; can filling; removal of air under vacuum; sealing/closing, retorting/heat treatment; cooling; labelling and packing. The vegetable may be canned whole, diced, puréed, as juice and so on.

On-line Simplified Methods for Enzyme Activity Check

Peroxidase Test:

(a) *Solutions:* In order to check the peroxidase activity two solutions have to be prepared:
   - 1% guaiacol in alcohol solution (1 g guaiacol is dissolved in about 50 cm^3 of 96% ethylic alcohol and then this preparation is brought to 100 cm° with the same solvent);
   - peroxide solution 0.3% (1 cm^3 perhydrol is brought to 100 cm^3 with distilled water.

(b) *Sampling:* From various parts of the material samples are taken (about 20-30 pieces, etc.); the material is then crushed in a laboratory bowl in order to obtain an average sample.

(c) *Check:* Prom the average sample, 10-20 g of material is taken in a medium capacity test tube; on this sample are poured: 20 cm^3 distilled water; 1 cm^3 of 1% guaiacol solution; 1.6 cm^3 of peroxide solution.

The contents of the test tube is shaken well. The gradual appearance of a weak pink colour indicates an incomplete peroxidase inactivation - reaction slightly positive.

If there are no tissue colour modifications after 5 minutes, the reaction is negative and the enzymes have been inactivated.

As an orientative check it is also possible to simply pour a few drops of 1% guaiacol solution and 0.3% peroxide solution directly on blanched and crushed vegetables. A rapid and intensive brown-reddish tissue colouring indicates a high peroxidase activity (positive reaction).

Catalase Test: In order to identify the catalase enzyme activity, 2 g of dehydrated vegetables are well crushed and mixed with about 20 cm^3 of distilled water. After 15 min softening, 0.5 cm^3 of a 0.5% or 1% peroxide solution is poured on prepared vegetables. In the presence of catalase, a strong oxygen generation is observed for about 2-3 minutes.

These tests are of a paramount importance in order to determine the vegetable blanching treatments (temperature and time); incomplete enzyme inactivation has a negative effect on finished product quality. For cabbage catalase inactivation by blanching is sufficient; blanching further to peroxidase inactivation would have negative effects on product quality and even complete browning. For all other vegetables and for potatoes, both tests MUST be negative, for catalase and for peroxidase.

Fresh Vegetable Storage

The vegetables can be stored, in some specific natural conditions, in fresh state, that is without significant modifications of their initial organoleptic properties. Fresh vegetable storage can be short term; this was briefly covered under temporary storage before processing. Also fresh vegetable storage can be long term during the cold season in some countries and in this case it is an important method for vegetable preservation in the natural state. In order to assure preservation in long term storage, it is necessary to reduce respiration and transpiration intensity to a minimum possible; this can be achieved by:

a. maintenance of as low a temperature as possible (down to 0° C),
b. air relative humidity increased up to 85-95 % and
c. CO_2 percentage in air related to the vegetable species.

Vegetables for storage must conform to following conditions: they must be of one of the autumn or winter type variety; be at edible maturity without going past this stage; be harvested during dry days; be protected from rain, sun heat or wind; be in a sound state and clean from soil; be undamaged. From the time of harvest and during all the period of their storage vegetables are subject to respiration and transpiration and this is on account of their reserve substances and water content. The more the intensity of these two natural processes are reduced, the longer sound storage time will be and the more losses will be reduced.

For this reason, vegetables have to be handled and transported as soon as possible in the storage conditions (optimal temperature and air relative humidity for the given species). Even in these optimal conditions storage will generate losses in weight which are variable and depend upon the species.

Some optimal storage conditions are shown in table.

Optimal Conditions for Fresh Vegetable Storage

Vegetables	Storage conditions	
	Temperature, °C	Relative humidity, %
Potatoes	+1...+3	85-90
Carrots	0 ... +1	90-95
Onions	0 ... +1	75-85
Leeks	0 ... +0.5	85-90
Cabbage	-1 ... 0	90-97
Garlic	0 ... +1	85-90
Beets	0 ... +1	90-95

7

Existing Post Harvest Technologies

Post Harvest technologies which greatly influence the level of Post Harvest losses and the quality of produce include grading, packaging, Pre-cooling, storage and transportation. Some products also require one or more of the following treatments: trimming, cleaning, curing, disease or insect control, waxing, and ripening. A general description of those technologies adopted in the United States is available in Kader's (1992) book. More detailed but less up-to-date information is also seen in books by Ryall and Lipton (1979), Ryall and Pentzer (1982), and Pantastico (1975). Storage requirements for a wide variety of products are summarized by Hardenburg et al. (1986).

Special requirements for each major product are often published in monographs, such as that for citrus fruits by Wardowski et al. (1986), and for banana by Stover and Simmonds (1987). Kader (1992) also provides long lists of scientific journals, journals with reviews or abstracts and other periodicals which contain up-to-date information of scientific findings and new technologies related to Post Harvest physiology and handling. This Bulletin is not intended to make a comprehensive review of literature, but will give a brief summary of current technology in common use.

Grading

Essentially all fruits and vegetables sold in modern markets are graded and sized. Sophisticated marketing systems require precise grading standards for each kind of product. More primitive markets may not use written grade standards, but the products are sorted and sized to some extent. Typical grading facilities in large packinghouses include dumpers and conveyors.

Produce is graded by human eyes and hands, while moving along conveyor belts or rollers. "Electric eyes" are sometimes used to sort produce by colour. In small-scale packing operations, one or a few grading tables may be enough. Dumping, conveying and grasping can cause mechanical injury to some products. Equipment should have a smooth, soft surface, and dumping and grading operations should be gentle, to minimize injuries.

Many products are sized according to their weight. Automated weight sizers of various capacity are used in packinghouses. Round or nearly round fruits are often sized according to their diameter, using automated chain or roller sizers or hand carried ring sizers. An inefficient sizing operation can also cause significant injuries.

Packaging

There are two very different types of packaging. The first is when produce is packed in containers for transportation and wholesale. The second is when produce is packed into small retail units.

Ideal containers for packing fruits and vegetables should have the following attributes. They are easy to handle, they provide good protection from mechanical damage, they have adequate ventilation, and they are convenient for merchandising (*i.e.* they can easily carry printed information and advertising about the product etc.). They should also be inexpensive, and easily degradable or recyclable. Many kinds of containers have been used, but the "ideal" is yet to be found. Users often put economic considerations first in selecting containers. Fancy containers such as fibreboard boxes, or wooden or plastic crates, are often used for high-value products. Inexpensive containers such as bamboo baskets or nylon net sacs are used for low-priced produce.

Methods of packing can affect the stability of products in the container during shipping, and influence how much the container protects their quality. In fibreboard boxes, for example, delicate and high-priced products are often packed in trays, while other products are simply all put in the box together.

Prepackaging or consumer packaging generally provides additional protection for the products. It is also convenient for retailers as well as customers, and therefore adds value to produce. However, overuse of non-biodegradable plastic trays and wrapping materials, as often seen in modern supermarkets, creates an extra burden of waste disposal and damages the environment.

Pre-cooling

Good temperature management is the most effective way to reduce Post Harvest losses and preserve the quality of fruits and vegetables. Desirable storage and transportation temperatures for major fruits and vegetables have been identified and published. Temperatives which are low, but not low enough to cause chilling-injury, slow down physiological activity and hence the rates of senescence of the products. Low temperatures also reduce microbial growth rates and the rate of decay. Products harvested from hot fields often carry field heat and have high rates of respiration.

Rapid removal of field heat by Pre-cooling is so effective in quality preservation that this procedure is widely used for highly perishable fruits and vegetables. Currently used Pre-cooling methods include room cooling, forced-air cooling, watercooling, vacuum cooling, and package icing. Cooling equipment and technologies for large-scale operations in the United States are well described in Kader's (1992) book. Smaller capacity equipment can be made using the same principles. Liu (1991) has made a brief summary of cooling principles and methods.

Room cooling is a relatively simple method which needs only a refrigerated room with adequate cooling capacity. The produce is packed in containers which are loosely stacked in the cooling room, leaving enough space between containers for each one to be exposed to circulating cold air. The rate of cooling is rather slow compared to other methods of cooling, because the heat inside each container needs to be transferred to the surface of the container by means of conduction before being carried away by the refrigerated air. It may take hours or even days to cool a product, depending on what kind of product it is, the size and nature of the container, and the temperature and velocity of the circulating air.

Forced-air cooling is a more rapid way of using air to cool produce. Cold air is forced to flow through the inside of each container, so that it carries away heat directly from the surface of the produce rather than from the surface of the container. The air flow is produced by creating a pressure difference between the two perforated sides of each container. The containers are stacked inside a covered tunnel with an exhaust fan at one end. Highly perishable and high-value products such as grapes, strawberries and raspberries may be cooled in less than an hour using this method.

Water cooling (also known as hydrocooling), is a rapid and less expensive method. Produce is exposed to cold water by means of showering or dipping. The required cooling time is often a matter of minutes. However, not all kinds of products tolerate hydrocooling. Hydrocooled products inevitably have a wet surface, which may encourage decay in some kinds of produce.

Vacuum cooling is the most efficient way to cool leafy vegetables, particularly headed ones such as head lettuce, cabbage and Chinese cabbage. The produce is placed inside a vacuum tube in which air pressure is reduced. When the pressure is lowered to 4.6 mm Hg, water "boils" off at 0°C from all over the leaf surface. The boiling effect draws heat for vaporization, and hence cools the produce. The cooling time is usually in the order of 20 to 30 minutes. Unfortunately, the equipment needed for vacuum cooling is very expensive, and may not be a good choice for small-scale farming systems.

Package-icing or top-icing is the simplest way of cooling. Adding crushed ice, flake-ice or slurry of ice in containers can cool the produce. However, this method is not suitable for produce which is very sensitive to ice-cold temperatures. Cooling by ice also inevitably wets both the produce and container, and generates water which needs to be drained. Fancy equipment is used in large-scale package-icing operations in the United States (Kader 1992), but many vegetable packers in Taiwan simply add ice flakes to each container, either by hand or by shovel.

Storage

Many horticultural crops have a relatively short harvesting season. Storage is needed to extend the marketing period. Various storage methods have been used on a commercial scale (Liu 1991). Air-cooled common storage houses are often used, or underground or cave storage using natural cold air. Storage humidity is sometimes regulated by controlled ventilation and dehumidifiers. Refrigerated storage (cold storage) controls temperature and humidity precisely by mechanical means. Controlled atmosphere (CA) storage controls the concentrations of oxygen and carbon dioxide, in addition to temperature and humidity.

Modified atmosphere (MA) storage also controls oxygen and carbon dioxide concentrations, although not as precisely as in CA, by using semipermeable polymeric films. Ethylene may be scrubbed for products responsive to it, regardless of the storage system used. A good control of temperature, humidity and atmospheric composition maximizes the storage life span of a product.

Air-cooled Common Storage

This is widely used for storing horticultural products, particularly those which have good keeping quality even without a precise low temperature. However, its use is generally limited to cool seasons in temperate and sub-tropical regions, or high altitude areas where there are low ambient temperatures at night.

An ideal storage room is adequately insulated and has a good ventilation control system which pulls cool air in at night and keeps warm air out during the day. Booth and Shaw (1981) and Hallee and Hunter (undated) have given comprehensive descriptions of how to construct and ventilate air-cooled common storage, but this technology has not been adopted in many parts of the world. For instance, hundreds of thousands of tons of citrus fruits are stored in common storage in Taiwan each winter, but ideal common storage facilities are still at their trial stage. Lack of technical knowledge, and the need for a large investment to build new facilities, delay adoption of this new technology.

Large commercial-scale storage of horticultural products in caves and underground cellars are seen only in northern and central China. The major crops stored are apples, pears and citrus fruits. Underground or cave storage often provides better temperature control than above-ground air-cooled storage. This method is most useful in areas where there is a long, cold winter, a dry climate, and a thick layer of fine heavy soil. The construction and management technologies developed in China do not yet seem to have spread to other countries. Some modification of Chinese technologies to suit different climatic and soil conditions might well make this technology useful in many parts of the world.

SA Storage (Controlled Atmosphere)

Commercial application of CA storage is limited to only a few crops, apples and pears being the most popular ones. It is not used for other crops because the benefit is too slight to cover the cost. The technologies involved (Bartsch and Blanpied 1984) are complicated and sophisticated. The cost of building, facilities, and management for CA storage is considerably higher than for refrigerated storage. It should not be recommended for any crop without a thorough cost and benefit analysis.

MA Storage

This is a much simpler method than CA. Theoretically, it is possible for MA storage to generate a result similar to that of CA. However, practical experience seems to show that MA storage is usually less reliable and less effective than CA, in terms of extending storage life and preserving quality. Its biggest advantages are that it is easy to use and not expensive. Therefore, commercial trials and application are becoming more common. Although many research papers have been published on trials using different polymeric films for wrapping or bagging different products, many new experiments are still conducted each year to try and get better results. This method is unlikely to replace CA for crops which are stored in large quantities over long periods, but it is often useful for small-scale and short-period storage of products which benefit from an atmosphere with a low oxygen and high carbon dioxide content.

Transportation

Inland transportation of horticultural crops is usually by rail or way truck. Overseas transportation is by sea or by air. A limited amount of high-valued produce is sometimes transported overland by air. The basic requirements for conditions during transportation are similar to those needed for storage, including proper control of temperature and humidity and adequate ventilation (except in the case of MA). In addition, the produce should be immobilized by proper packaging and stacking, to avoid excessive movement or vibration. Vibration and impact during transportation may cause severe bruising or other types of mechanical injury.

CA containers are not used much for transporting fruit and vegetables. Only a small number of CA containers are used for long-distance shipping of highly perishable, high-value products in the United States and for international trade. Refrigerated containers and trailers are more often used for long-distance shipping, whether by sea, rail or truck. Modern technologies for the refrigerated transportation of horticultural produce have been well described by McGregor (1987).

Shipping by refrigerated trucks is not only convenient, but also effective in preserving the quality of product. However, both the initial investment and the operating costs are very high. Another possibility is insulated or properly ventilated trailer trucks. Pre-cooled products can be transported by well insulated non-refrigerated trucks for up to several hours without any significant rise in product temperature. There are considerable cost savings without any sacrifice of quality if trucks are only insulated, rather than refrigerated, for short-distance shipping. If the product is not Pre-cooled or if the shipping distance is long, a ventilated truck is a better choice than an insulated truck without ventilation and without refrigeration. Ventilation alone does not usually provide a uniform cool temperature, but it may help dissipate excessive field heat and respiration heat, and thus avoid high temperature injury.

Difficulties in Applying Western Technology

Modern Post Harvest technologies developed over the past twenty or thirty years are now widely used in North America and Europe. Japan is the only Asian country which uses technologies of comparable sophistication.

Although there are many Asian Post Harvest horticulturists who have been trained in the United States, Europe, or Japan and returned to their home countries, technical improvement in many Asian countries has been slow. There are several reasons why Western technologies cannot be applied quickly in Asia. Firstly, Western technologies use sophisticated equipment and facilities which are too large for the small-scale farming systems of Asia. There is some small-scale equipment manufactured in Europe or Japan, but it is too expensive for most Asian countries. Secondly, advantages of applying Western technologies are less obvious in traditional marketing systems than in the supermarkets which dominate retail in western countries.

Large capacity coolers, waxers, grading and packing line facilities, etc. manufactured in the United States are designed for use in large packinghouses. Asian farmers not only have small-scale operations, but also grow diverse crops. Their products are either packed by the growers themselves, or in small packingsheds operated by farmers' cooperatives. The volume each grower or cooperative packs each day is too small for large equipment. Although smaller equipment is often made in European countries and Japan, it is always sold at high prices.

The manufacturers have good reason to ask high prices, because the number of units sold is so small. Reputable manufacturers of farm machinery are reluctant to compromise the quality of their products for the sake of lower prices, so they make only sophisticated, expensive equipment. Since user countries cannot even afford to buy the first unit for trial, it is not possible for them to order many units at once for a bargain price. Modern supermarkets have refrigerated display cabinets and temporary cold storage. They can keep produce cool continuously, provided it has been properly cooled before arrival. In contrast, traditional markets do not usually have refrigeration.

They make every effort to sell produce quickly to reduce display time and avoid quality loss. Even if the produce has been cooled before and during shipping, the "cold chain" breaks at the retail markets and hence reduces the benefit of cooling. Even worse, many consumers mistakenly believe that cold produce has been stored, while warm produce means it is "fresh". This misconception further discourages cooling, which is the most effective way to prevent rapid deterioration of the produce.

Modification of High-Technology

If the capacity of a packing shed is small, only small-scale equipment is needed. If a country cannot afford to buy expensive equipment, cheaper equipment should be developed. In order to lower the price, it may be necessary to sacrifice quality in terms of appearance, convenience, mobility, durability and/or user friendliness. Sacrificing one or more of these attributes might be acceptable at the initial stage of development, provided that the equipment still functions well. In order to encourage users to purchase, the equipment must be sold at an affordable price. Therefore, sophisticated equipment and facilities developed in Western countries may need substantial modification to suit local needs in Asian countries.

This is best done in the user countries. The equipment and facilities needed for Post Harvest handling of horticultural crops are relatively simple. The principles involved have been widely published, and are readily available to farm machinery specialists, while few or no patents are involved. The fundamental need for developing modified equipment of this kind is good cooperation among Post Harvest horticulturists, farm machinery specialists and end users, plus sufficient support from government policies. Several Asian countries have made significant progress in this area in recent years. For example, locally made small-scale forced-air coolers, crushed-ice makers, and small refrigerated storage facilities are often seen at cooperative shipping points in Taiwan today.

Integration: The Key to Success

A typical route for a horticultural product to follow from harvest to arrival in the consumer's hands includes grading, packaging, transportation, wholesaling and retailing. Additional processes which might be added en route are trimming, curing, Pre-cooling, storage, waxing, disease and insect control treatments, and prepackaging. Improper procedures at any step en route may ruin all the efforts made during other steps to preserve the quality of the product. In other words, the final quality of the product is often determined by the worst procedure, which acts as a limiting factor.

Since no Post Harvest technology can remedy quality which is already damaged, a mistake at one point may not be compensated by extra care in other steps. Therefore, an integrated programme covering all Post Harvest procedures is necessary to guarantee success. For instance, if three steps in the Post Harvest handling system need to be improved, they all have to be improved simultaneously. Improving only one or two steps may not be any improvement at all.

Taiwan had at least two bad experiences in the past, when it made fragmentary effort to improve Post Harvest technology. One was to import two hydrocoolers from the United States in the 1970s for cooling vegetables produced in central Taiwan. The government subsidized more than half the cost of the equipment, but the users (Farmers' Associations) still could not justify the operational costs for the very limited benefits from using the coolers.

There were no suitable transportation and storage facilities to protect hydrocooled products at that time. The products rewarmed rapidly. Pre-cooled and then rewarmed wet vegetables often decay faster than those which have not been cooled at all. The ultimate fate of the hycrocoolers was to be dismantled and sold as waste metal.

The second failure was the subsidizing of small refrigerated trucks to transport fresh vegetables from eastern Taiwan to Taipei in the 1980s. Since no Pre-cooling facilities were available at the shipping point, the warm vegetables remained warm or became warmer inside the "refrigerated" truck, because the cooling capacity of the truck was enough to keep cool vegetables from warming but not enough to precool warm products.

After these costly lessons of failure, Post Harvest workers realized that improvement programmes should be integrated rather than fragmentary. The people who formulate and supervise a programme must have a comprehensive knowledge of the subject. They should understand all procedures involved, and be able to identify all problems and choose remedial strategies.

The first successful programme in Taiwan began with vegetables in the suburbs. Harvested vegetables were trimmed, hydrocooled, prepackaged, stored in refrigerated rooms until needed, and then shipped to supermarkets in Taipei by insulated small trucks, with or without refrigeration. The hydrocoolers, refrigerated storage, and shipping trucks used were all small and were locally made.

Encouraged by the favorable response of growers and consumers, the programme was extended to major vegetable production areas in central and southern Taiwan. As the volume of vegetables handled under the project increased and shipping distance lengthened, larger coolers, refrigeration storage rooms and shipping trucks came into use. In order to avoid wetting vegetables, forced-air cooling gradually replaced hydrocooling for some kinds of vegetables and fruit. More and more shipping points adopted the new technology, and Pre-cooled products were not only shipped to supermarkets, but also to large institutional users, and sometimes wholesale markets. Post Harvest losses were reduced, and product value increased.

In order to provide shippers, extension staff and marketing personnel with the essential technical knowledge, a number of short training courses were offered. At the beginning of the programme, Government agencies begged users to try the new technology. Nowadays, it is the users who beg government agencies for newer technology and more technical advice.

Research into New Technologies

Existing technologies for cooling, storage, and transportation of horticultural crops are generally adequate at present. New research should aim at improving equipment, facilities and methods, to make them more efficient and less costly. Applied research for solving specific problems is likely to be more fruitful than basic research in developing countries. Different crops often need different Post Harvest treatments. Post Harvest horticulturists must establish the requirements for each major crop. Even if optimum conditions and best treatments are known, it is often worthwhile to study the crop's response to less than optimum conditions, since the best treatment is not necessarily the most economic one. For instance, a study was made of optimum storage conditions for Taiwan citrus fruits. This was followed by experiments in refrigerated storage, improved common storage, and "traditional" common storage. Cave storage is also going to be studied. When these experiments are completed, we shall be able to make recommendations for growers, depending on their specific location and the duration of the storage.

Since improved handling methods often rely heavily on improvements in equipment and facilities, close cooperation between Post Harvest horticulturists and agricultural engineers is necessary. An area which is urgently in need of new technology is Post Harvest disease and insect control. Effective fungicides for Post Harvest application have become fewer and fewer, since many previously used fungicides are now banned for fear of health hazards. Control of Post Harvest decay must rely on new, safe chemical treatments and practical physical methods, in addition to improved sanitary conditions. Recent findings, such as the use of acetic acid fumigation, (Moyls et al. 1996) and UV-C light (Wilson et al. 1997) indicate the feasibility of finding new chemical and physical methods. Many countries wish to export their quality products, but have difficulty in fulfilling the quarantine requirements imposed by importing countries. Current quarantine procedures are either chemical or physical (Paull and Armstrong 1994). Many tropical products cannot tolerate any form of cold treatment. Heat treatment is expensive, and often damages the quality of products. Irradiation is a possibility, but still needs further development. New and safe chemical treatments are yet to be found.

Coordination and Cooperation

More about Taiwan's Experience

Although our past experiences of making improving Post Harvest handling of horticultural crops in Taiwan have not all been successful, significant progress has been made in recent years. In the early stages, even a small success could be realized only after many trials and frustrations. Projects which began in the 1970s ended in almost total failure, because every programme tackled only part of the Post Harvest system programmes were also technically unsound, according to current knowledge. More projects, were launched in the 1980s and some small progress made, but many projects still failed. Eventually the stage was reached in the late 1980s when there were enough high-caliber Post Harvest scientists, while both industry and government agencies were aware of the importance of Post Harvest handling. Taiwan was then able to carry out more effective programmes.

Recent progress has been based on good cooperation between Post Harvest scientists, government coordinators, extension staff, and decision makers at shipping points and terminal markets. Full coordination and cooperation were made possible after repeated discussions and debates. Once the major funding agencies were convinced that an integrated national programme should replace individual fragmentary small projects, the programme really took off.

This type of integrated programmes quickly generated fruitful results. Each year we were able to see some improvement in the quality of fresh produce at the markets, and growers and shippers began to enjoy the benefits of reduced Post Harvest losses and increased profit margins for higher value-added products.

Several steps were crucial in changing programmes from initial failure to recent success:

- A few high-level Post Harvest scientists were trained to serve as programme leaders, while more than a hundred middle-level technical workers were trained to implement projects.
- Key persons were organized into a strong working group.
- Problems, priorities, and programmes were discussed, to generate a consensus within the group.
- Fragmentary projects were integrated into a national programme.

Post Harvest Handling of Horticulture Produce

Horticultural crops not only provide us with nutritional and healthy foods, but also generate a cash income to growers. Appropriate production practices, careful harvesting and proper packaging, storage and transport all contribute to the good produce quality. Once a crop is harvested it is impossible to improve its quality.

The horticultural crops, because of their high moisture content are inherently more liable to deteriorate especially under tropical conditions. Moreover, they are biologically active and carry out transpiration, respiration, ripening and other biochemical activities, which deteriorate the quality of the produce.

Losses during Post Harvest operations due to improper storage and handling are enormous and can range from 10-40 percent. Post Harvest losses can occur in the field, in packing areas, in storage, during transportation and in the wholesale and retail market. Severe losses occur because of poor facilities, lack of know-how, poor management, market dysfunction or simply the carelessness of farmers. Proper storage conditions, temperature and humidity are needed to lengthen the storage life and maintain quality once the crop has been cooled to the optimum storage temperature.

Causes of Post Harvest Losses

Physiological and Biochemical Aspects

The quality of the harvested fruits and vegetables depend on the condition of growth as well as physiological and biochemical changes they undergo after harvest. Fruits and vegetable cells are still alive after harvest and continue their physiological activity. The Post Harvest quality and storage life of fruits appear to be controlled by the maturity. If the fruits are harvested at a proper stage of maturity the quality of the fruits is excellent. Poor quality and uneven ripening are due to early harvesting and late harvesting which results in extremely poor shelf life.

Respiration plays a very significant role in the Post Harvest life of the fruits. In most of the fruits, the rate of respiration increases rapidly with ripening. The sudden upsurge in respiration is called the 'climacteric rise', which is considered to be the turning point in the life of the fruit. After this the senescence and deterioration of the fruit begin. The fruits such as banana, papaya, mango, guava, jackfruit, fig, sapota, etc. belong to the category of climacteric fruits. While litchi, pineapple, grapes, pomegranate, lemon, orange, lime, etc. belong to the non-climacteric group. To extend the Post Harvest life of the fruits its respiration rate should be reduced as far as possible. Thus an understanding of the factors, which influence the rate of respiration, is indispensable to Post Harvest technologies for manipulating the storage behaviour of fruits.

Mechanical Injury

Owing to their tender texture and high moisture content, fresh fruits and vegetables are very susceptible to mechanical injury. Poor handling, unsuitable containers, improper packaging and transportation can easily cause bruising, cutting, breaking, impact wounding and other forms of injury.

Parasitic Diseases

High Post Harvest losses are caused by the invasion of fungi, bacteria, insects and other organisms. Microorganisms attack fresh produce easily and spread quickly, because the produce does not have much of a natural defence mechanism and has plenty of nutrients and moisture to support microbial growth. Post Harvest decay control is becoming a more difficult task, because the number of pesticides available is falling rapidly as consumer concern for food safety increases.

Post Harvest Technologies

Pre-cooling: Good temperature management is the most effective way to reduce Post Harvest losses and preserve the quality of fruits and vegetables. Products harvested from hot fields often carry field heat and have high rates of respiration. Rapid removal of field heat by Pre-cooling is so effective in quality preservation that this procedure is widely used for highly perishable fruits and vegetables. Currently used Pre-cooling methods include room cooling, forced-air cooling, water cooling, vacuum cooling and package icing

Room cooling is a relatively simple method, which needs only a refrigerated room with adequate cooling capacity. The produce is packed in containers, which are loosely stacked in the cooling room, leaving enough space between containers for each one to be exposed to circulating cold air.

The rate of cooling is rather slow compared to other methods of cooling, because the heat inside each container needs to be transferred to the surface of the container by means of conduction before being carried away by the refrigerated air. It may take hours or even days to cool a product, depending on what kind of product it is, the size and nature of the container, and the temperature and velocity of the circulating air.

Forced-air cooling is a more rapid way of using air to cool produce. Cold air is forced to flow through the inside of each container, so that it carries away heat directly from the surface of the produce rather than from the surface of the container. The airflow is produced by creating a pressure difference between the two perforated sides of each container. The containers are stacked inside a covered tunnel with an exhaust fan at one end. Highly perishable and high-value products such as grapes, strawberries and raspberries may be cooled in less than an hour using this method.

Hydro cooling is a rapid and less expensive method. Produce is exposed to cold water by means of showering or dipping. The required cooling time is often a matter of minutes. However, not all kinds of products tolerate hydro cooling. Hydro cooled products inevitably have a wet surface, which may encourage decay in some kinds of produce.

Vacuum cooling is the most efficient way to cool leafy vegetables, particularly headed ones such as head lettuce, cabbage and Chinese cabbage. The produce is placed inside a vacuum tube in which air pressure is reduced. When the pressure is lowered to 4.6 mm Hg, water "boils" off at 0°C from all over the leaf surface. The boiling effect draws heat for vaporization, and hence cools the produce. The cooling time is usually in the order of 20 to 30 minutes. Unfortunately, the equipment needed for vacuum cooling is very expensive, and may not be a good choice for small-scale farming systems. Ice bank cooler is a new development in refrigeration with positive ventilation. In this system ice cool air is passed through the boxes containing horticultural produce. This facilitates quicker cooling and large amount of heat is removed in a relatively shorter period. The store maintains a temperature of 0.5 - 0.8°C and relative humidity of 98 percent. Package icing or top icing is the simplest way of cooling. Adding crushed ice, flake-ice or slurry of ice in containers can cool the produce. However, this method is not suitable for produce, which is very sensitive to ice-cold temperatures. Cooling by ice also inevitably wets both the produce and container, and generates water, which needs to be drained.

Sanitation

Sanitation is of great concern to produce handlers, not only to protect produce against Post Harvest diseases, but also to protect consumers from food borne illnesses. E.coli 157:H_7, Salmonella, Chryptosporidium, Hepatitis and Cyclospera are among the disease causing organisms that have been transferred via fresh fruits and vegetables. Use of a disinfectant in wash water can help to prevent both Post Harvest diseases and food borne illnesses.

Chlorine in the form of a sodium hypochlorite solution or as a dry powdered calcium hypochlorite can be used in hydro-cooling or wash water as a disinfectant. For the majority of vegetables, chlorine in wash water should be maintained in the range of 75-150 ppm (parts per million).

The antimicrobial form, hypochlorous acid, is mostly available in water with a neutral pH (6.5 to 7.5). Organic growers must use chlorine with caution, as it is classified as a restricted material.

Ozonation is another technology that can be used to sanitize produce. A naturally occurring molecule, ozone is a powerful disinfectant. Fruit and vegetable growers have begun using it in dump tanks as well, where it can be thousands of times more effective than chlorine. Ozone not only kills whatever food borne pathogens might be present, it also destroys microbes responsible for spoilage. A basic system consists of an ozone generator, a monitor to gauge and adjust the levels of ozone being produced and a device to dissolve the ozone gas into the water.

Hydrogen peroxide can also be used as a disinfectant. Concentrations of 0.5% or less are effective for inhibiting development of Post Harvest decay caused by a number of fungi. Hydrogen peroxide has a low toxicity rating and is generally recognized as having little potential for environmental damage.

Presizing and Storage

For many commodities fruits below a certain size are eliminated manually or mechanically by presizing belt. Undersized fruits are diverted for processing. The sorting process eliminates cull, overripe, misshapen and otherwise defective fruit and separates produce by colour, maturity and ripeness classes.

Waxing

Food grade waxes are commonly applied to replace some of the natural waxes removed in the washing and cleaning operations to reduce water loss and to improve appearance. It also provides protection against decay organisms. Waxing may be done after grading and fungicides may be added to the wax. Application of wax and Post Harvest fungicides must be indicated on each container where the refrigerated storage facilities are not available Protective skin coating with wax is one of the methods for increasing the storage life of fresh fruits.

Packaging

Packaging of fresh fruits and vegetables has a great significance in reducing the wastage. Packaging provides protection from physical damage during storage, transportation and marketing. There are variety of packages, packaging materials and inserts available.

There are two types of packaging. The first is when produce is packed in containers for transportation and wholesale. The second is when produce is packed into small retail units. Ideal containers for packing fruits and vegetables should have the following attributes. They are easy to handle, they provide good protection from mechanical damage, they have adequate ventilation and they are convenient for merchandising. They should also be inexpensive and easily degradable or recyclable. Many kinds of containers have been used but the "ideal" is yet to be found. Users often put economic considerations first is selecting containers. Fancy containers such as fibreboard boxes or wooden or plastic crates, are often used for high-value products. Inexpensive containers such as bamboo baskets or nylon net sacs are used for low-priced produce. Methods of packaging can affect the stability of products in the container during shipping, and influence how much the container protects their quality. In fibreboard boxes, for example, delicate and high-priced products are often packed in trays, while other products are simply put in the box in groups. Prepackaging or consumer packaging generally provides additional protection for the products. It is also convenient for retailers as well as customers, and therefore adds value to produce. However, overuse of non-biodegradable plastic trays and wrapping materials, as often seen in modern supermarkets, which creates an extra burden of waste disposal and damages the environment.

Factors Affecting Storage Life

Relative Humidity

Transpiration rates (water loss from produce) are determine by the moisture content of the air, which is usually expressed as relative humidity. At high relative humidity, produce maintains salable weight, appearance, nutritional quality and flavour, while wilting, softening and juiciness are reduced. Leafy vegetables with high surface-to-volume ratios; injured produce and immature fruits and vegetables have higher transpiration rates. High temperatures, low relative humidity and high air velocity increase transpiration rates.

Relative humidity needs to be monitored and controlled in storage. Control can be achieved by a variety of methods:

1. Operating a humidifier in the storage area.
2. Regulating air movement and ventilation in relation to storage room load.
3. Maintaining refrigeration coil temperature within the storage room.
4. Using moisture barriers in the insulation of the storage room or transport vehicle.
5. Wetting the storage room floor.
6. Using crushed ice to pack produce for shipment.
7. Sprinkling leafy vegetables, cool-season root vegetables and immature fruits and vegetables with water.

Temperature

Respiration and metabolic rates are directly related to room temperatures within a given range. The higher the rate of respiration, the faster the produce deteriorates. Lower temperatures reduce respiration rates and the ripening and senescence processes, which prolong the storage life of fruits and vegetables. Low temperatures also slow the growth of pathogenic fungi, which cause spoilage of fruits and vegetables in storage.

Producers should give special care and attention to proper storage conditions for produce with high to extremely high respiration rates, as these crops will deteriorate much more quickly.

It is impossible to make a single recommendation for cool storage of all fruits and vegetables. Climate of the area where the crop originated, the plant part, the season of harvest and crop maturity at harvest are important factors in determining the optimum temperature. A general rule for vegetables is that cool-season crops should be stored at cooler temperatures (0 to 1.7°C) and warm-season crops should be stored at warmer temperatures (7 to 13°C).

Freezing Injury

Temperatures that are too low can be just as damaging as those too high. Freezing will occur in all commodities below 0°C. Whether injury occurs depends on the commodity. Some can be repeatedly frozen and thawed without damage, while others are ruined by one freezing. Injury from freezing temperatures can appear in plant tissues as loss of rigidity, softening and water soaking. Injury can be reduced if the produce is allowed to warm up slowly to optimum storage temperatures and if it is not handled during the thawing period. Injured produce should be marketed immediately, as freezing shortens its storage life.

Chilling Injury

Fruits and vegetables that require warmer storage temperatures (4.5 to 13°C) can be damaged if they are subjected to near-freezing temperatures (0°C). Cooler temperatures interfere with normal metabolic processes. Injury symptoms are varied and often do not develop until the produce has been returned to warmer temperatures for several days. Besides physical damage, chilled produce is often more susceptible to disease infection.

Ethylene

Ethylene, a natural hormone produced by some fruits as they ripen, promotes additional ripening of produce exposed to it. The old age saying that one bad apple spoils the whole bushel is true.

The damaged or diseased fruits produce high levels of ethylene and stimulate the other apples to ripen too quickly. As the fruits ripen, they become more susceptible to disease. Ethylene "producers" should not be stored with fruits, vegetables, or flowers that are sensitive to it. The result could be loss of quality, reduced shelf life and specific symptoms of injury. Ethylene producers include apples, apricots, avocados, ripening bananas, honeydew melons, papayas, peaches, pears, plums and tomatoes.

Storage Facilities

Crops that require different storage conditions will need three different storage facilities.

- Cold Storage (temperatures 0 to 2.2°C)
- Cool Storage (temperatures 4.5 to 13°F)
- Warmer storage (temperatures 13 to 15.6°F)

A recording thermometer can be helpful in determining whether storage facilities are maintaining ideal conditions and are not fluctuating. A maximum/minimum thermometer could be substituted. Relative humidity also should be monitored with a hygrometer.

Controlling and monitoring temperature and relative humidity will enable a grower to maintain optimum storage conditions for maximum storage life of the crop and to minimize crop damage from chilling, freezing or high temperature injuries and water loss from the crop.

Air-Cooled Common Storage

This is widely used for storing horticultural products, particularly those that have good keeping quality even without a precise low temperature. However, its use is generally limited to cool seasons in temperate and sub-tropical regions, or high altitude areas where there are low ambient temperatures at night. An ideal storage room is adequately insulated and has a good ventilation control system, which draws cool air inside during night and keeps warm air out during the day.

Refrigerated Storage

Refrigerated storage is a well-established technology widely used for storing horticultural crops all over the world. Its application is limited only by cost and benefit considerations. Essentially, all crops can benefit by being stored at a suitable low temperature, which extends the storage life and preserves quality.

Many horticultural crops have storage life spans ranging from less than one month to several months when refrigerated. Therefore, refrigerated storage can be used continuously only if different crops with different harvesting seasons can share the facility. There are other important reasons why this method is not used in many tropical and sub-tropical countries, where refrigeration is needed most. The initial investment cost is too high and its energy consumption too large for many countries.

Hydrobaric Storage

In this system the horticultural produce is kept in a vacuum-tight and refrigerated container and the air is evacuated by vacuum pump to achieve desired low pressure. The low pressure retards ripening by decreasing respiration. However this is more expensive method.

Transportation

Inland transportation of horticultural crops is usually by rail or by truck. Overseas transportation is by sea or by air. A limited amount of high-valued produce is sometimes transported overland by air. The basic requirements for conditions during transportation are proper control of temperature and humidity and adequate ventilation. In addition, the produce should be immobilized by proper packaging and stacking, to avoid excessive movement or vibration. Vibration and impact during transportation may cause severe bruising or other types of mechanical injury. Refrigerated containers and trailers are more often used for long distance shipping, whether by sea, rail or truck.

Shipping by refrigerated trucks is not only convenient, but also effective in preserving the quality of product. However, both the initial investment and the operating costs are very high. Another possibility is insulated or properly ventilated trailer trucks. Pre-cooled products can be transported through well-insulated non-refrigerated trucks for up to several hours without any significant rise in product temperature. There are considerable cost savings without any sacrifice of quality if trucks are only insulated, rather than refrigerated, for short-distance shipping. If the product is not Pre-cooled or if the shipping distance is long, a ventilated truck is a better choice than an insulated truck without ventilation and without refrigeration. Ventilation alone does not usually provide a uniform cool temperature, but it may help dissipate excessive field heat and respiration heat, and thus avoid high temperature injury.

Post Harvest Technology of Horticulture

In developing countries agriculture is the mainstay of the economy. As such, it should be no surprise that agricultural industries and related activities can account for a considerable proportion of their output. Of the various types of activities that can be termed as agriculturally based, fruit and vegetable processing are among the most important.

Both established and planned fruit and vegetable processing projects aim at solving a very clearly identified development problem. This is that due to insufficient demand, weak infrastructure, poor transportation and perishable nature of the crops, the grower sustains substantial losses. During the Post Harvest glut, the loss is considerable and often some of the produce has to be fed to animals or allowed to rot.

Even established fruit and vegetable canning factories or small/medium scale processing centres suffer huge loss due to erratic supplies. The grower may like to sell his produce in the open market directly to the consumer, or the produce may not be of high enough quality to process even though it might be good enough for the table. This means that processing capacities will be seriously underexploited.

The main objective of fruit and vegetable processing is to supply wholesome, safe, nutritious and acceptable food to consumers throughout the year.

Fruit and vegetable processing projects also aim to replace imported products like squash, yams, tomato sauces, pickles, etc., besides earning foreign exchange by exporting finished or semi-processed products.

The fruit and vegetable processing activities have been set up, or have to be established in developing countries for one or other of the following reasons:

- diversification of the economy, in order to reduce present dependence on one export commodity;
- government industrialisation policy;
- reduction of imports and meeting export demands;
- stimulate agricultural production by obtaining marketable products;
- generate both rural and urban employment;

- reduce fruit and vegetable losses;
- improve farmers' nutrition by allowing them to consume their own processed fruit and vegetables during the off-season;
- generate new sources of income for farmers/artisans;
- develop new value-added products.

What Fruit and Vegetables can be Processed?

Practically any fruit and vegetable can be processed, but some important factors which determine whether it is worthwhile are:

a. The demand for a particular fruit or vegetable in the processed form;

b. The quality of the raw material, *i.e.* whether it can withstand processing;

c. Regular supplies of the raw material.

For example, a particular variety of fruit which may be excellent to eat fresh is not necessarily good for processing. Processing requires frequent handling, high temperature and pressure.

Many of the ordinary table varieties of tomatoes, for instance, are not suitable for making paste or other processed products. A particular mango or pineapple may be very tasty eaten fresh, but when it goes to the processing centre it may fail to stand up to the processing requirements due to variations in its quality, size, maturity, variety and so on.

Even when a variety can be processed, it is not suitable unless large and regular supplies are made available. An important processing centre or a factory cannot be planned just to rely on seasonal gluts; although it can take care of the gluts it will not run economically unless regular supplies are guaranteed.

To operate a fruit and vegetable processing centre efficiently it is of utmost importance to pre-organise growth, collection and transport of suitable raw material, either on the nucleus farm basis or using outgrowers.

Processing Planning

The secret of a well planned fruit and vegetable processing centre is that it must be designed to operate for as many months of the year as possible. This means the facilities, the buildings, the material handling and the equipment itself must be interlinked and coordinated properly to allow as many products as possible to be handled at the same time, and yet the equipment must be versatile enough to be able to handle many products without major alterations.

A typical processing centre or factory should process four or five types of fruits harvested at different times of the year and two or three vegetables. This processing unit must also be capable of handling dried/dehydrated finished products, juices, pickles, tomato juice, ketchup and paste, jams, jellies and marmalades, semi-processed fruit products. Advanced planning is necessary to process a large range of products in varied weather and temperature conditions, each requiring a special set of manufacturing and packaging formulae. The end result of the efforts should be a well-managed processing unit with lower initial investment.

A unit which is sensibly laid out and where one requirement co-relates to another, with a sound costing analysis, leads to an integrated operation.

Instead of over-sophisticated machinery, a sensible simple processing unit may be required when planned production is not very large and is geared mainly to meet the demand of the domestic market.

Location

The basic objective is to choose the location which minimises the average production cost, including transport and handling.

It is an advantage, all other things being equal, to locate a processing unit near the fresh raw material supply. It is a necessity for proper handling of the perishable raw materials, it allows the processing unit to allow the product to reach its best stage of maturation and lessens injury from handling and deterioration from changes during long transportation after harvesting.

An adequate supply of good water, availability of manpower, proximity to rail or road transport facilities and adequate markets are other important requirements.

Processing Systems

a. *Small-Scale Processing:* This is done by small-scale farmers for personal subsistence or for sale in nearby markets. In this system, processing requires little investment: however, it is time consuming and tedious. Until recently, small-scale processing satisfied the needs of rural and urban populations. However, with the rising rates of population and urbanisation growth and their more diversified food demands, there is need for more processed and diversified types of food.

b. *Intermediate-Scale Processing:* In this scale of processing, a group of small-scale processors pool their resources. This can also be done by individuals. Processing is based on the technology used by small-scale processors with differences in the type and capacity of equipment used. The raw materials are usually grown by the processors themselves or are purchased on contract from other farmers. These operations are usually located on the production site of in order to assure raw materials availability and reduce cost of transport. This system of processing can provide quantities of processed products to urban areas.

c. *Large-Scale Processing:* Processing in this system is highly mechanised and requires a substantial supply of raw materials for economical operation. This system requires a large capital investment and high technical and managerial skills. Because of the high demand for foods in recent years many large-scale factories were established in developing countries. Some succeeded, but the majority failed, especially in West Africa. Most of the failures were related to high labour inputs and relatively high cost, lack of managerial skills, high cost and supply instability of raw materials and changing governmental policies. Perhaps the most important reason for failure was lack of adequate quantity and regularity of raw material supply to factories. Despite the failure of these commercial operations, they should be able to succeed with better planning and management, along with the undertaking of more in-depth feasibility studies.

It can be concluded that all three types of processing systems have a place in developing countries to complement crop production to meet food demand. Historically, however, small and intermediate scale processing proved to be more successful than large-scale processing in developing countries.

Choice of Processing Technologies for Developing Countries

FAO maintains (in FAO, 1992c), that the basis for choosing a processing technology for developing countries ought to be to combine labour, material resources and capital so that not only the type and quantity of goods and services produced are taken into account, but also the distribution of their benefits and the prospects of overall growth. These should include:

a. Increasing farmer/artisan income by the full utilisation of available indigenous raw material and local manufacturing of part or all processing equipment;

b. Cutting production costs by better utilisation of local natural resources (solar energy) and reducing transport costs;

c. Generating and distributing income by decentralising processing activities and involving different beneficiaries in processing activities (investors, newly employed, farmers and small-scale industry);

d. maximising national output by reducing capital expenditure and royalty payments, more effectively developing balance-of-payments deficits through minimising imports (equipment, packing material, additives), and maximising export-oriented production;

e. Maximising availability of consumer goods by maximisation of high-quality, standard processed produce for internal and export markets, reducing Post Harvest losses, giving added value to indigenous crops and increasing the volume and quality of agricultural output.

Knowledge and control of the means of production, local manufacturing of processing equipment and development of appropriate/new technologies and more suitable raw material for processing must all be better researched.

Decentralisation of activities must be maintained and coordinated. The introduction of more sophisticated processing equipment and packaging material must be subordinated to internal and export marketing references.

Choosing a technology solely to maximise profits can actually work against true development. Choice should also be based on a solid, long-term market opportunity to ensure viability.

The internal market should be given greater consideration, safeguarded and supported.

Training courses, at all levels, in processing and preservation of indigenous crops, must be expanded.

Chemical Composition and Nutritional Aspects

General Properties

Fruit and vegetables have many similarities with respect to their compositions, methods of cultivation and harvesting, storage properties and processing. In fact, many vegetables may be considered fruit in the true botanical sense. Botanically, fruits are those portions of the plant which house seeds. Therefore such items as tomatoes, cucumbers, eggplant, peppers, and others would be classified as fruits on this basis.

However, the important distinction between fruit and vegetables has come to be made on an usage basis. Those plant items that are generally eaten with the main course of a meal are considered to be vegetables. Those that are commonly eaten as dessert are considered fruits. That is the distinction made by the food processor, certain marketing laws and the consuming public, and this distinction will be followed in this document.

Vegetables are derived from various parts of plants and it is sometimes useful to associate different vegetables with the parts of the plant they represent since this provides clues to some of the characteristics we may expect in these items.

Fruit as a dessert item, is the mature ovaries of plants with their seeds. The edible portion of most fruit is the fleshy part of the pericarp or vessel surrounding the seeds. Fruit in general is acidic and sugary. They commonly are grouped into several major divisions, depending principally upon botanical structure, chemical composition and climatic requirements.

Berries are fruit which are generally small and quite fragile. Grapes are also physically fragile and grow in clusters. Melons, on the other hand, are large and have a tough outer rind. Drupes (stone fruit) contain single pits and include such items as apricots, cherries, peaches and plums. Pomes contain many pits, and are represented by apples, quince and pears.

Citrus fruit like oranges, grapefruit and lemons are high in citric acid. Tropical and subtropical fruits include bananas, dates, figs, pineapples, mangoes, and others which require warm climates, but exclude the separate group of citrus fruits.

The compositions of representative vegetables and fruits in comparison with a few of the cereal grains are seen in Table.

Typical Percentage Composition of Foods from Plant Origin Percentage Composition- Edible Portion

Food	*Carbo-hydrate*	*Protein*	*Fat*	*Ash*	*Water*
Cereals					
wheat flour, white	73.9	10.5	1.9	1.7	12
rice, milled, white	78.9	6.7	0.7	0.7	13
maize, whole grain	72.9	9.5	4.3	1.3	12
Earth vegetables					
potatoes, white	18.9	2.0	0.1	1.0	78
sweet potatoes	27.3	1.3	0.4	1.0	70
Vegetables					
carrots	9.1	1.1	0.2	1.0	88.6
radishes	4.2	1.1	0.1	0.9	93.7
asparagus	4.1	2.1	0.2	0.7	92.9
beans, snap, green	7.6	2.4	0.2	0.7	89.1
peas, fresh	17.0	6.7	0.4	0.9	75.0
lettuce	2.8	1.3	0.2	0.9	94.8
Fruit					
banana	24.0	1.3	0.4	0.8	73.5
orange	11.3	0.9	0.2	0.5	87.1
apple	15.0	0.3	0.4	0.3	84.0
strawberries	8.3	0.8	0.5	0.5	89.9

Source: *Anon. (196O)*

Compositions of vegetables and fruit not only vary for a given kind in according to botanical variety, cultivation practices, and weather, but change with the degree of maturity prior to harvest, and the condition of ripeness, which is progressive after harvest and is further influenced by storage conditions. Nevertheless, some generalisations can be made.

Most fresh vegetables and fruit are high in water content, low in protein, and low in fat. In these cases water contents will generally be greater than 70% and frequently greater than 85%.

Commonly protein content will not be greater than 3.5% or fat content greater than 0.5 %. Exceptions exist in the case of dates and raisins which are substantially lower in moisture but cannot be considered fresh in the same sense as other fruit. Legumes such as peas and certain beans are higher in protein; a few vegetables such as sweet corn which are slightly higher in fat and avocados which are substantially higher in fat. Vegetables and fruit are important sources of both digestible and indigestible carbohydrates. The digestible carbohydrates are present largely in the form of sugars and starches while indigestible cellulose provides roughage which is important to normal digestion. Fruit and vegetables are also important sources of minerals and certain vitamins, especially vitamins A and C. The precursors of vitamin A, including beta-carotene and certain other carotenoids, are to be found particularly in the yellow-orange fruit and vegetables and in the green leafy vegetables.

Citrus fruit are excellent sources of vitamin C, as are green leafy vegetables and tomatoes. Potatoes also provide an important source of vitamin C for the diets of many countries. This is not so much due to the level of vitamin C in potatoes which is not especially high but rather to the large quantities of potatoes consumed.

Chemical Composition

Water: Vegetal cells contain important quantities of water. Water plays a vital role in the evolution and reproduction cycle and in physiological processes. It has effects on the storage period length and on the consumption of tissue reserve substances.

In vegetal cells, water is present in following forms:

- bound water or dilution water which is present in the cell and forms true solutions with mineral or organic substances;
- colloidal bound water which is present in the membrane, cytoplasm and nucleus and acts as a swelling agent for these colloidal structure substances; it is very difficult to remove during drying/dehydration processes;
- constitution water, directly bound on the chemical component molecules and which is also removed with difficulty.

Vegetables contain generally 90-96% water while for fruit normal water content is between 80 and 90%.

Mineral substances: Mineral substances are present as salts of organic or inorganic acids or as complex organic combinations (chlorophyll, lecithin, etc.); they are in many cases dissolved in cellular juice.

Vegetables are more rich in mineral substances as compared with fruits. The mineral substance content is normally between 0.60 and 1.80% and more than 60 elements are present; the major elements are: K, Na, Ca, Mg, Fe, Mn, Al, P. Cl, S.

Among the vegetables which are especially rich in mineral substances are: spinach, carrots, cabbage and tomatoes. Mineral rich fruit includes: strawberries, cherries, peaches and raspberries. Important quantities of potassium (K) and absence of sodium chloride (NaCl) give a high dietetic value to fruit and to their processed products. Phosphorus is supplied mainly by vegetables.

Vegetables usually contain more calcium than fruit; green beans, cabbage, onions and beans contain more than 0.1% calcium. The calcium/phosphorus or Ca/P ratio is essential for calcium fixation in the human body; this value is considered normal at 0.7 for adults and at 1.0 for children. Some fruit are important for their Ca/P ratio above 1.0: pears, lemons, oranges and some temperate climate mountain fruits and wild berries. Even if its content in the human body is very low, iron (Fe) has an important role as a constituent of haemoglobin. Main iron sources are apples and spinach. Salts from fruit have a basic reaction; for this reason fruit consumption facilitates the neutralisation of noxious uric acid reactions and contributes to the acid-basic equilibrium in the blood.

Carbohydrates: Carbohydrates are the main component of fruit and vegetables and represent more than 90% of their dry matter. From an energy point of view carbohydrates represent the most valuable of the food components; daily adult intake should contain about 500 g carbohydrates. Carbohydrates play a major role in biological systems and in foods. They are produced by the process of photosynthesis in green plants. They may serve as structural components as in the case of cellulose; they may be stored as energy reserves as in the case of starch in plants; they may function as essential components of nucleic acids as in the case of ribose; and as components of vitamins such as ribose and riboflavin. Carbohydrates can be oxidised to furnish energy, and glucose in the blood is a ready source of energy for the human body. Fermentation of carbohydrates by yeast and other microorganisms can yield carbon dioxide, alcohol, organic acids and other compounds.

Some properties of sugars. Sugars such as glucose, fructose, maltose and sucrose all share the following characteristics in varying degrees, related to fruit and vegetable technology:

- they supply energy for nutrition;
- they are readily fermented by micro-organisms;
- in high concentrations they prevent the growth of micro-organisms, so they may be used as a preservative;
- on heating they darken in colour or caramelise;
- some of them combine with proteins to give dark colours known as the browning reaction.

Some properties of starches:

- They provide a reserve energy source in plants and supply energy in nutrition;
- they occur in seeds and tubers as characteristic starch granules.

Some properties of celluloses and hemicelluloses:

- They are abundant in the plant kingdom and act primarily as supporting structures in the plant tissues;
- they are insoluble in cold and hot water;
- they are not digested by man and so do not yield energy for nutrition;
- the fibre in food which produces necessary roughage is largely cellulose.

Some properties of pectins and carbohydrate gums.

- Pectins are common in fruits and vegetables and are gumlike (they are found in and between cell walls) and help hold the plant cells together;
- pectins in colloidal solution contribute to viscosity of the tomato paste;
- pectins in solution form gels when sugar and acid are added; this is the basis of jelly manufacture.

Fats: Generally fruit and vegetables contain very low level of fats, below 0.5%. However, significant quantities are found in nuts (55%), apricot kernel (40%), grapes seeds (16%), apple seeds (20%) and tomato seeds (18%).

Organic Acids

Fruit contains natural acids, such as citric acid in oranges and lemons, malic acid of apples, and tartaric acid of grapes. These acids give the fruits tartness and slow down bacterial spoilage. We deliberately ferment some foods with desirable bacteria to produce acids and this give the food flavour and keeping quality. Examples are fermentation of cabbage to produce lactic acid and yield sauerkraut and fermentation of apple juice to produce first alcohol and then acetic acid to obtain vinegar.

Organic acids influence the colour of foods since many plant pigments are natural pH indicators. With respect to bacterial spoilage, a most important contribution of organic acids is in lowering a food's pH. Under anaerobic conditions and slightly above a pH of 4.6, Clostridium botulinum can grow and produce lethal toxins. This hazard is absent from foods high in organic acids resulting in a pH of 4.6 and less.

Acidity and sugars are two main elements which determine the taste of fruit. The sugar/acid ratio is very often used in order to give a technological characterisation of fruits and of some vegetables.

Nitrogen-containing Substances

These substances are found in plants as different combinations: proteins, amino acids, amides, amines, nitrates, etc. Vegetables contain between 1.0 and 5.5 % while in fruit nitrogen-containing substances are less than 1% in most cases.

Among nitrogen containing substances the most important are proteins; they have a colloidal structure and, by heating, their water solution above 50°C an one-way reaction makes them insoluble. This behaviour has to be taken into account in heat processing of fruits and vegetables. From a biological point of view vegetal proteins are less valuable then animal ones because in their composition all essential amino-acids are not present.

Vitamins

Vitamins are defined as organic materials which must be supplied to the human body in small amounts apart from the essential amino-acids or fatty acids. Vitamins function as enzyme systems which facilitate the metabolism of proteins, carbohydrates and fats but there is growing evidence that their roles in maintaining health may extend yet further. The vitamins are conveniently divided into two major groups, those that are fat-soluble and those that are water-soluble. Fat-soluble vitamins are A, D, E and K. Their absorption by the body depends upon the normal absorption of fat from the diet. Water-soluble vitamins include vitamin C and several members of the vitamin B complex.

Vitamin-A or Retinol

This vitamin is found as such only in animal materials - meat, milk, eggs and the like. Plants contain no vitamin A but contain its precursor, beta-carotene. Man needs either vitamin A or beta-carotene which he can easily convert to vitamin A. Beta-carotene is found in the orange and yellow vegetables as well as the green leafy vegetables, mainly carrots, squash, sweet potatoes, spinach and kale. A deficiency of vitamin A leads to night blindness, failure of normal bone and tooth development in the young and diseases of epithelial cells and membrane of the nose, throat and eyes which decrease the body's resistance to infection.

Vitamin C

Vitamin C is the anti-scurvy vitamin. Lack of it causes fragile capillary walls, easy bleeding of the gums, loosening of teeth and bone joint diseases. It is necessary for the normal formation of the protein collagen, which is an important constituent of skin and connective tissue. Like vitamin E, vitamin C favours the absorption of iron. Vitamin C, also known as ascorbic acid, is easily destroyed by oxidation especially at high temperatures and is the vitamin most easily lost during processing, storage and cooking.

Excellent sources of vitamin C are citrus fruits, tomatoes, cabbage and green peppers. Potatoes also are a fair source (although the content of vitamin C is relatively low) because we consume large quantities of potatoes.

Enzymes

Enzymes are biological catalysts that promote most of the biochemical reactions which occur in vegetable cells.

Some properties of enzymes important in fruit and vegetable technology are the following:

- in living fruit and vegetables enzymes control the reactions associated with ripening;
- after harvest, unless destroyed by heat, chemicals or some other means, enzymes continue the ripening process, in many cases to the point of spoilage - such as soft melons or overripe bananas;
- because enzymes enter into a vast number of biochemical reactions in fruits and vegetable, they may be responsible for changes in flavour, colour, texture and nutritional properties;
- the heating processes in fruit and vegetables manufacturing/processing are designed not only to destroy micro-organisms but also to deactivate enzymes and so improve the fruit and vegetables' storage stability.

Enzymes have an optimal temperature - around +50°C where their activity is at maximum. Heating beyond this optimal temperature deactivates the enzyme. Activity of each enzyme is also characterised by an optimal pH. In fruit and vegetable storage and processing the most important roles are played by the enzymes classes of hydrolases (lipase, invertase, tannase, chlorophylase, amylase, cellulase) and oxidoreductases (peroxidase, tyrosinase, catalase, ascorbinase, polyphenoloxidase).

Turgidity and Texture

The range of textures that are encountered in fresh and cooked vegetables and fruit is indeed great, and to a large extent can be explained in terms of changes in specific cellular components. Since plants tissues generally contain more than two-thirds water, the relationships between these components and water further determine textural differences.

Cell Turgidity: Quite apart from other contributing factors, the state of turgidity, determined by osmotic forces, plays a paramount role in the texture of fruit and vegetables. The cell walls of plant tissues have varying degrees of elasticity and are largely permeable to water and ions as well as to small molecules. The membranes of the living protoplast are semi-permeable, that is they allow passage of water but are selective with respect to transfer of dissolved and suspended materials. The cell vacuoles contain most of the water in plant cells and sugars, acids, salts, amino acids, some water-soluble pigments and vitamins, and other low molecular weight constituents are dissolved in this water. In the living plant, water taken up by the roots passes through the cell walls and membranes into the cytoplasm of the protoplasts and into the vacuoles to establish a state of osmotic equilibrium within the cells.

The osmotic pressure within the cell vacuoles and within the protoplasts pushes the protoplasts against the cell walls and causes them to stretch slightly in accordance with their elastic properties. This is the situation in the growing plant and the harvested live fruit or vegetable which is responsible for desired plumpness, succulence, and much of the crispness. When plant tissues are damaged or killed by storage, freezing, cooking, or other causes, an important major change that results is denaturation of the proteins of cell membranes resulting in the loss of perm-selectivity. Without perm-selectivity the state of osmotic pressure in cell vacuoles and protoplasts cannot exist, and water and dissolved substances are free to diffuse out of the cells and leave the remaining tissue in a soft and wilted condition.

Other Factors Affecting Texture. The existence of a high degree of turgidity in live fruit and vegetables or whether a relative state of flabbiness develops from loss of osmotic pressure as well as final texture depends on several cell constituents.

Cellulose, Hemicellulose, and Lignin. Cell walls in young plants are very thin and are composed largely of cellulose. As the plant ages cell walls tend to thicken and become higher in hemicellulose and in lignin. These materials are fibrous and tough and are not significantly softened by cooking.

Pectic Substances. The complex polymers of sugar acid derivatives include pectin and closely related substances. The cement-like substance found especially in the middle lamella which helps hold plant cells to one another is a water-insoluble pectic substance.

On mild hydrolysis it yields water-soluble pectin which can form gels or viscous colloidal suspensions with sugar and acid. Certain water-soluble pectic substances also react with metal ions, particularly calcium, to form water-insoluble salts such as calcium pectates. The various pectic substances may influence texture of vegetables and fruits in several ways.

When vegetables or fruit are cooked, some of the water-insoluble pectic substance is hydrolysed into water-soluble pectin. This results in a degree of cell separation in the tissues and contributes to tenderness. Since many fruits and vegetables are somewhat acidic and contain sugars the soluble pectin also tends to form colloidal suspensions which will thicken the juice or pulp of these products.

Fruit and vegetables also contain a natural enzyme which can further hydrolyse pectin to the point where the pectin loses much of its gel forming property. This enzyme is known as pectin methyl esterase. Materials such as tomato juice or tomato paste will contain both pectin and pectin methyl esterase.

If freshly prepared tomato juice or paste is allowed to stand the original viscosity gradually decreases due to the action of pectin methyl esterase on pectin gel.

This can be prevented if the tomato products are quickly heated to a temperature of about 82°C (180 F°) to deactivate the pectin methyl esterase liberated from broken cells before it has a chance to hydrolyse the pectin. Such a treatment is commonly practiced in the manufacture of tomato juice products. This is known as the "hot-break process" and yields products of high viscosity. In contrast, where low viscosity products are desired no heat is used and enzyme activity is allowed to proceed. This is "cold-break" process. After sufficient decrease in viscosity is achieved the product can be heat treated, as in canning, to preserve it for long term storage. It is often also desirable to firm the texture of fruit and vegetables, especially when products are normally softened by processing. In this case advantage is taken of the reaction between soluble pectic substances and calcium ions which form calcium pectates. These calcium pectates are water insoluble and when they are produced within the tissues of fruit and vegetables they increase structural rigidity. Thus, it is common commercial practice to add low levels of calcium salts to tomatoes, apples, and other vegetables and fruits prior to canning or freezing.

Sources of Colour and Colour Changes

In addition to a great range of textures, much of the interest that fruits and vegetables add to our diets is due to their delightful and variable colours. The pigments and colour precursors of fruit and vegetables occur for the most part in the cellular plastic inclusions such as the chloroplasts and other chromoplasts, and to a lesser extent dissolved in fat droplets or water within the cell protoplast and vacuoles.

These pigments are classified into four major groups which include the chlorophylls, carotenoids, anthocyanins, and anthoanthins. Pigments belonging to the latter two groups also are referred to as flavonoids, and include the tannins.

The Chlorophylls. The chlorophylls are contained mainly within the chloroplasts and have a primary role in the photosynthetic production of carbohydrates from carbon dioxide and water. The bright green colour of leaves and other parts of plants is largely due to the oilsoluble chlorophylls, which in nature are bound to protein molecules in highly organised complexes. When the plant cells are killed by ageing, processing, or cooking, the protein of these complexes is denatured and the chlorophyll may be released. Such chlorophyll is highly unstable and rapidly changes in colour to olive green and brown. This colour change is believed to be due to the conversion of chlorophyll to the compound pheophytin.

Conversion to pheophytin is favoured by acid pH but does not occur readily under alkaline conditions. For this reason peas, beans, spinach, and other green vegetables which tend to lose their bright green colours on heating can be largely protected against such colour changes by the addition of sodium bicarbonate or other alkali to the cooking or canning water.

However, this practice is not looked upon favourably nor used commercially because alkaline pH also has a softening effect on cellulose and vegetable texture and also destroys vitamin C and thiamin at cooking temperatures.

The Carotenoids: Pigments belonging to this group are fat-soluble and range in colour from yellow through orange to red. They often occur along with the chlorophylls in the chloroplasts, but also are present in other chromoplasts and may occur free in fat droplets. Important carotenoids include the orange carotenes of carrot, maize, apricot, peach, citrus fruits, and squash; the red lycopene of tomato, watermelon, and apricot; the yellow-orange xanthophyll of maize, peach, paprika and squash; and the yellow-orange crocetin of the spice saffron. These and other carotenoids seldom occur singly within plant cells.

A major importance of some of the carotenoids is their relationship to vitamin A. A molecule of orange beta-carotene is converted into two molecules of colourless vitamin A in the animal body. Other carotenoids like alpha-carotene, gamma-carotene, and cryptoxanthin also are precursors of vitamin A, but because of minor differences in chemical structure one molecule of each of these yields only one molecule of vitamin A.

In food processing the carotenoids are fairly resistant to heat, changes in pH, and water leaching since they are fat-soluble. However, they are very sensitive to oxidation, which results in both colour loss and destruction of vitamin A activity.

The Flavonoids: Pigments and colour precursors belonging to this class are water-soluble and commonly are present in the juices of fruit and vegetables. The flavonoids include the purple, blue, and red anthocyanins of grapes, berries, plump, eggplant, and cherry; the yellow anthoxanthins of light coloured fruit and vegetables such as apple, onion, potato, and cauliflower, and the colourless catechins and leucoanthocyanins which are food tannins and are found in apples, grapes, tea, and other plant tissues. These colourless tannin compounds are easily converted to brown pigments upon reaction with metal ions. Properties of the anthocyanins include a shifting of colours with pH. Thus many of the anthocyanins which are violet or blue in alkaline media become red upon addition of acid.

Cooking of beets with vinegar tends to shift the colour from a purplish red to a brighter red, while alkaline water can influence the colour of red fruits and vegetables toward violet and gray-blue. The anthocyanins also tend toward the violet and blue hues upon reaction with metal ions, which is one reason for lacquering the inside of metal cans when the true colour of anthocyanin-containing fruits and vegetables is to be preserved.

The water-soluble property of anthocyanins also results in easy leaching of these pigments from cut fruit and vegetables during processing and cooking.

The yellow anthoxanthins also are pH sensitive tending toward a deeper yellow in alkaline media. Thus potatoes or apples become somewhat yellow when cooked in water with a pH of 8 or higher, which is common in many areas. Acidification of the water to pH 6 or lower favours a whiter colour.

The colourless tannin compounds upon reaction with metal ions form a range of dark coloured complexes which may be red, brown, green, grey, or black. The various shades of these coloured complexes depend upon the particular tannin, the specific metal ion, pH, concentration of the complex, and other factors not yet fully understood. Water-soluble tannins appear in the juices squeezed from grapes, apples, and other fruits as well as the brews from extraction of tea and coffee. The colour and clarity of tea are influenced by the hardness and pH of the brewing water. Alkaline waters that contain calcium and magnesium favour the formation of dark brown tannin complexes which precipitate when the tea is cooled.

If acid in the form of lemon juice is added to such tea its colour lightens and the precipitate tends to dissolve. Iron from equipment or from pitted tin cans has caused a number of unexpected colours to develop in products containing tannins, such as coffee, cocoa and foods flavoured with these. The tannins are also important because they have an astringency which influences flavour and contributes body to such beverages as tea, wine, apple cider, etc.

Fruit and vegetables are in a live state after harvest. Continued respiration gives off carbon dioxide, moisture, and heat which influence storage, packaging, and refrigeration requirements. Continued transpiration adds to moisture evolved and further influences packaging requirements. Further activities of fruit and vegetables, before and after harvest, include changes in carbohydrates, pectins, organic acids, and the effects these have on various quality attributes of the products.

As for changes in carbohydrates, few generalizations can be given with respect to starches and sugars. In some plant products sugars quickly decrease and starch increases in amount soon after harvest. This is the case for ripe sweet corn which can suffer flavour and texture quality losses in a very few hours after harvest.

Unripe fruit, in contrast, is frequently high in starch and low in sugars. Continued ripening after harvest generally results in a decrease in starch and a increase in sugars as in the case of apples and pears. However, this does not necessarily mean that the starch is the source of the newly formed sugars. Further, the courses of change in starch and sugars are markedly influenced by Post Harvest storage temperatures. Thus potatoes stored below about 10 C° (50 F°) continue to build up high levels of sugars, while the same potatoes stored above 10 C° do not.

This property is used to help the dehydration process in potato storage. Here potatoes should have a low reducing sugar content so as to minimise Maillard browning reactions during drying and subsequent storage of the dried product. In this case potatoes are stored above 10°C prior to being further processed.

After harvest the pectin changes in fruit and vegetables are more predictable. Generally there is decrease in water-insoluble pectic substance and a corresponding increase in water soluble pectin. This contributes to the gradual softening of fruits and vegetables during storage and ripening. Further breakdown of water-soluble pectin by pectin methyl esterase also occurs. The organic acids of fruit generally decreases during storage and ripening. This occurs in apples and pears and is especially important in the case of oranges. Oranges have a long ripening period on the tree and time of picking is largely determined by degree of acidity and sugar content which have major effects upon juice quality. It is important to note that the reduction of acid content on ripening influences more than just the tartness of fruit. Since many of the plant pigments are sensitive to acid, fruit colour would be expected to change. Additionally, the viscosity of pectin gel is affected by acid and sugar contents, both of which change with ripening.

Stability of Nutrients

One of the principal responsibilities of the food scientist and food technologist is to preserve food nutrients through all phases of food acquisition, processing, storage, and preparation. The key is in the specific sensitivities of the various nutrients, the principles of which are illustrated.

Specific Sensitivity and Stability of Nutrients

Nutrient	*Neutral* p^{H7}	*Acid* < p^{H7}	*Alkaline* > p^{H7}	*Air or Oxygen*	*Light*	*Heat*	*Cooking Losses, Range*
			Vitamins				
Vitamin A	S	U	S	U	U	U	0-40
Ascorbic acid(C)	U	S	U	U	U	U	0-100
Biotin	S	S	S	S	S	U	0-60
Carotenes	S	U	S	U	U	U	0-30
(pro A)							0-5
Choline	S	S	S	U	S	S	0-10
Cobalamin	S	S	S	U	U	S	
(B_{12})							0-40
Vitamin D	S		U	U	U	U	0-10
Essential fatty acids	S	S	U	U	U	S	
Folic acid	U	U	S	U	U	U	0-100
Inositol	S	S	S	S	S	U	0-95
Vitamin K	S	U	U	S	U	S	0-5
Niacin (PP)	S	S	S	S	S	S	0-75
Pantbothenic acid	S	U	U	S	S	U	0-50
p-Amino Benzoic acid	S	S	S	U	S	S	0-5
Vitamin B_6	S	S	S	S	U	U	0-40

Riboflavin (B_2)	S	S	U	S	U	U	0-75
Tbiamin (B_1)	U	S	U	U	S	U	0-80
Tocopherols	S	S	S	U	U	U	0-55

**Source:* Harris and Karmas, 1975

(U = Unstable; S = Stable)

This shows the stability of vitamins, essential amino acids, and minerals to acid, air, light, and heat, and gives an indication of possible cooking losses. Vitamin A is highly sensitive to acid, air, light and heat; vitamin C to alkalinity, air, light and heat; vitamin D to alkalinity, air, light and heat; thiamin to alkalinity, air, and heat in alkaline solutions; etc. Cooking losses of some essential nutrients may be in excess of 75%. In modern food processing operations, however, losses are seldom in excess of 25%.

The ultimate nutritive value of a food results from the sum total of losses incurred throughout its history - from farmer to consumer. Nutrient value begins with genetics of the plant and animal. The farmland fertilization programme affects tissue composition of plants, and animals consuming these plants. The weather and degree of maturity at harvest affect tissue composition.

Storage conditions before processing affect vitamins and other nutrients. Washing, trimming, and heat treatments affect nutrient content. Canning, evaporating, drying, and freezing alter nutritional values, and the choices of times and temperatures in these operations frequently must be balanced between good bacterial destruction and minimum nutrient destruction.

Packaging and subsequent storage affect nutrients. One of the most important factors is the final preparation of the food in the home and the restaurant - the steam table can destroy much of what has been preserved through all prior manipulations.

Structural Features

The structural unit of the edible portion of most fruits and vegetables is the parenchyma cell. While parenchyma cells of different fruit and vegetables differ somewhat in gross size and appearance, all have essentially the same fundamental structure.

Parenchyma cells of plants differ from animal cells in that the actively metabolising protoplast portion of plant cells represents only a small fraction, of the order of five per cent, of the total cell volume. This protoplast is film-like and is pressed against the cell wall by the large water-filled central vacuole.

The protoplast has inner and outer semi-permeable membrane layers; the cytoplasm and its nucleus are held between them. The cytoplasm contains various inclusions, among them starch granules and plastics such as the chloroplasts and other pigment-containing chromoplasts. The cell wall, cellulose in nature, contributes rigidity to the parenchyma cell and limits the outer protoplasmic membrane. It is also the structure against which other parenchyma cells are cemented to form extensive three-dimensional tissue masses.

The layer between cell walls of adjacent parenchyma cells is referred to as the middle lamella, and is composed largely of pectic and polysaccharide cement-like materials. Air spaces also exist, especially at the angles formed where several cells come together. The relationships between these structures and their chemical compositions are further outlined below. The parenchyma cells will vary in size among plants but are quite large when compared to bacterial or yeast cells. The larger parenchyma cells may have volumes many thousand times greater than a typical bacterial cell.

There are additional types of cells other than parenchyma cells that make up the familiar structures of fruit and vegetables. These include various types of conducting cells which are tube-like and distribute water and salts throughout the plant.

Such cells produce fibrous structures toughened by the presence of cellulose and the wood-like substance lignin. Cellulose, lignin, and pectic substances also occur in specialised supporting cells which increase in importance as plants become older.

An important structural feature of all plants, including fruit and vegetables is protective tissue. This can take many forms but usually is made up of specialised parenchyma cells that are pressed compactly together to form a skin, peel or rind.

Surface cells of these protective structures on leaves, stems or fruit secrete waxy cutin and form a water impermeable cuticle. These surface tissues, especially on leaves and young stems will also contain numerous valve-like cellular structures, the stomata, through which moisture and gases can pass.

Deterioration Factors and their Control

Enzymic Changes

Enzymes which are endogenous to plant tissues can have undesirable or desirable consequences. Examples involving endogenous enzymes include (a) the Post Harvest senescence and spoilage of fruit and vegetables; (b) oxidation of phenolic substances in plant tissues by phenolase (leading to browning); (c) sugar-starch conversion in plant tissues by amylases; (d) Post Harvest demethylation of pectic substances in plant tissues (leading to softening of plant tissues during ripening, and firming of plant tissues during processing).

The major factors useful in controlling enzyme activity are: temperature, water activity, pH, chemicals which can inhibit enzyme action, alteration of substrates, alteration of products and pre-processing control.

Chemical Changes

Sensory Quality: The two major chemical changes which occur during the processing and storage of foods and lead to a deterioration in sensory quality are lipid oxidation and non-enzymatic browning. Chemical reactions are also responsible for changes in the colour and flavour of foods during processing and storage.

1. Lipid oxidation rate and course of reaction is influenced by light, local oxygen concentration, high temperature, the presence of catalysts (generally transition metals such as iron and copper) and water activity. Control of these factors can significantly reduce the extent of lipid oxidation in foods.
2. Non-enzymic browning is one of the major causes of deterioration which occurs during storage of dried and concentrated foods. The non-enzymic browning, or Maillard reaction, can be divided into three stages: (a) early Maillard reactions which are chemically well-defined steps without browning; (b) advanced Maillard reactions which lead to the formation of volatile or soluble substances; and (c) final Maillard reactions leading to insoluble brown polymers.

Colour Changes

Chlorophylls: Almost any type of food processing or storage causes some deterioration of the chlorophyll pigments. Phenophytinisation (with consequent formation of a dull olive-brown phenophytin) is the major change; this reaction is accelerated by heat and is acid catalysed.

Other reactions are also possible. For example, dehydrated products such as green peas and beans packed in clear glass containers undergo photo-oxidation and loss of desirable colour.

Anthocyanins: These are a group of more than 150 reddish water-soluble pigments that are very widespread in the plant kingdom. The rate of anthocyanin destruction is pH dependent, being greater at higher pH values. Of interest from a packaging point of view is the ability of some anthocyanins to form complexes with metals such as Al, Fe, Cu and Sn.

These complexes generally result in a change in the colour of the pigment (for example, red sour cherries react with tin to form a purple complex) and are therefore undesirable. Since metal packaging materials such as cans could be sources of these metals, they are usually coated with special organic linings to avoid these undesirable reactions.

Carotenoids: The carotenoids are a group of mainly lipid soluble compounds responsible for many of the yellow and red colours of plant and animal products. The main cause of carotenoid degradation in foods is oxidation. The mechanism of oxidation in processed foods is complex and depends on many factors. The pigments may auto-oxidise by reaction with atmospheric oxygen at rates dependent on light, heat and the presence of pro- and antioxidants.

Flavour Changes: In fruit and vegetables, enzymically generated compounds derived from long-chain fatty acids play an extremely important role in the formation of characteristic flavours. In addition, these types of reactions can lead to significant off-flavours. Enzyme-induced oxidative breakdown of unsaturated fatty acids occurs extensively in plant tissues and this yield characteristic aromas associated with some ripening fruits and disrupted tissues.

The permeability of packaging materials is of importance in retaining desirable volatile components within packages, or in permitting undesirable components to permeate through the package from the ambient atmosphere.

Nutritional Quality

The four major factors which affect nutrient degradation and can be controlled to varying extents by packaging are light, oxygen concentration, temperature and water activity. However, because of the diverse nature of the various nutrients as well as the chemical heterogeneity within each class of compounds and the complex interactions of the above variables, generalizations about nutrient degradation in foods will inevitably be broad ones.

Vitamins: Ascorbic acid is the most sensitive vitamin in foods, its stability varying markedly as a function of environmental conditions such as pH and the concentration of trace metal ions and oxygen. The nature of the packaging material can significantly affect the stability of ascorbic acid in foods. The effectiveness of the material as a barrier to moisture and oxygen as well as the chemical nature of the surface exposed to the food are important factors.

For example, problems of ascorbic acid instability in aseptically packaged fruit juices have been encountered because of oxygen permeability of the package and the oxygen dependence of the ascorbic acid degradation reaction.

Also, because of the preferential oxidation of metallic tin, citrus juices packaged in cans with a tin contact surface exhibit greater stability of ascorbic acid than those in enamelled cans or glass containers. The aerobic and anaerobic degradation reactions of ascorbic acid in reduced-moisture foods have been shown to be highly sensitive to water activity, the reaction rate increasing in an exponential fashion over the water activity range of 0.1-0.8.

Physical Changes

One major undesirable physical change in food powders is the absorption of moisture as a consequence of an inadequate barrier provided by the package; this results in caking. It can occur either as a result of a poor selection of packaging material in the first place, or failure of the package integrity during storage. In general, moisture absorption is associated with increased cohesiveness. Anti-caking agents are very fine powders of an inert chemical substance that are added to powders with much larger particle size in order to inhibit caking and improve flowability. Studies in onion powders showed that at ambient temperature, caking does not occur at water activities of less than about 0.4. At higher activities, however, (aw > 0.45) the observed time to caking is inversely proportional to water activity, and at these levels anti-caking agents are completely ineffective. It appears that while they reduce inter-particle attraction and interfere with the continuity of liquid bridges, they are unable to cover moisture sorption sites.

Biological Changes

Microbiological: Micro-organisms can make both desirable and undesirable changes to the quality of foods depending on whether or not they are introduced as an essential part of the food preservation process or arise unintentionally and subsequently grow to produce food spoilage.

The two major groups of micro-organisms found in foods are bacteria and fungi, the latter consisting of yeasts and moulds. Bacteria are generally the fastest growing, so that in conditions favourable to both, bacteria will usually outgrow fungi.

Foods are frequently classified on the basis of their stability as non-perishable, semi-perishable and perishable. For example, hermetically sealed and heat processed (*e.g.* canned) foods are generally regarded as non-perishable. However, they may become perishable under certain circumstances when an opportunity for recontamination is afforded following processing.

Such an opportunity may arise if the can seams are faulty, or if there is excessive corrosion resulting in internal gas formation and eventual bursting of the can. Spoilage may also take place when the canned food is stored at unusually high temperatures: thermophilic spore-forming bacteria may multiply, causing undesirable changes such as flat sour spoilage. Low moisture content foods such as dried fruit and vegetables are classified as semi-perishable. Frozen foods, though basically perishable, may be classified as semi-perishable provided that they are properly stored at freezer temperatures.

The majority of foods (*e.g.* meat and fish, milk, eggs and most fresh fruits and vegetables) are classified as perishable unless they have been processed in some way. Often, the only form of processing which such foods receive is to be packaged and kept under controlled temperature conditions. The species of micro-organisms which cause the spoilage of particular foods are influenced by two factors: a) the nature of the foods and b) their surroundings. These factors are referred to as intrinsic and extrinsic parameters.

The intrinsic parameters are an inherent part of the food: pH, aw, nutrient content, antimicrobial constituents and biological structures. The extrinsic parameters of foods are those properties of the storage environment that affect both the foods and their microorganisms. The growth rate of the micro-organisms responsible for spoilage primarily depends on these extrinsic parameters: temperature, relative humidity and gas compositions of the surrounding atmosphere. The protection of packaged food from contamination or attack by micro-organisms depends on the mechanical integrity of the package (*e.g.* the absence of breaks and seal imperfections), and on the resistance of the package to penetration by micro-organisms.

Metal cans which are retorted after filling can leak during cooling, admitting any microorganisms which may be present in the cooling water, even when the double seam is of a high quality. This fact is widely known in the canning industry and is the reason for the mandatory chlorination of cannery cooling water. Extensive studies on a variety of plastic films and metal foils have shown that microorganisms (including mounds, yeasts and bacteria) cannot penetrate these materials in the absence of pinholes. In practice, however, thin sheets of packaging materials such as aluminium and plastic do contain pinholes. There are several safeguards against the passage of micro-organisms through pinholes in films:

- because of surface tension effects, micro-organisms cannot pass through very small pinholes unless the micro-organisms are suspended in solutions containing wetting agents and the pressure outside the package is greater than that within;
- materials of packaging are generally used in thicknesses such that pinholes are very infrequent and small;
- for applications in which package integrity is essential (such as sterilisation of food in pouches), adequate test methods are available to assure freedom from bacterial recontamination.

Macrobiological

Insect Pests: Warm humid environments promote insect growth, although most insects will not breed if the temperature exceeds about 35 C° or falls below 10 C°. Also many insects cannot reproduce satisfactorily unless the moisture content of their food is greater than about 11%. The main categories of foods subject to pest attack are cereal grains and products derived from cereal grains, other seeds used as food (especially legumes), dairy products such as cheese and milk powders, dried fruits, dried and smoked meats and nuts.

As well as their possible health significance, the presence of insects and insect excrete in packaged foods may render products unsaleable, causing considerable economic loss, as well as reduction in nutritional quality, production of off-flavours and acceleration of decay processes due to creation of higher temperatures and moisture levels.

Early stages of infestation are often difficult to detect; however, infestation can generally be spotted not only by the presence of the insects themselves but also by the products of their activities such as webbing, clumped-together food particles and holes in packaging materials.

Unless plastic films are laminated with foil or paper, insects are able to penetrate most of them quite easily, the rate of penetration usually being directly related to film thickness. In general, thicker films are more resistant than thinner films, and oriented films tend to be more effective than cast films.

The looseness of the film has also been reported to be an important factor, loose films being more easily penetrated than tightly fitted films.

Generally, the penetration varies depending on the basic resin from which the film is made, on the combination of materials, on the package structure, and of the species and stage of insects involved. The relative resistance to insect penetration of some flexible packaging materials is as follows:

- excellent resistance: polycarbonate; poly-ethylene-terephthalate;
- good resistance: cellulose acetate; polyamide; polyethylene (0.254 mm); polypropylene (biaxially oriented); poly-vinyl-chloride (unplasticised);
- fair resistance: acrylonitrile; poly-tetra-fluoro-ethylene; polyethylene (0.123 mm);
- poor resistance: regenerated cellulose; corrugated paper board; kraft paper; polyethylene (0.0254-0.100 mm); paper/foil/polyethylene laminate pouch; poly-vinylchloride (plasticised).

Some simple methods for obtaining insect resistance of packaging materials are as following:

- select a film and a film thickness that are inherently resistant to insect penetration;
- use shrink film over-wraps to provide an additional barrier;
- seal carton flaps completely.

Rodents: Rats and mice carry disease-producing organisms on their feet and/or in their intestinal tracts and are known to harbour salmonella of sero types frequently associated with food-borne infections in humans.

In addition to the public health consequences of rodent populations in close proximity to humans, these animals also compete intensively with humans for food. Rats and mice gnaw to reach sources of food and drink and to keep their teeth short.

Their incisor teeth are so strong that rats have been known to gnaw through lead pipes and unhardened concrete, as well as sacks, wood and flexible packaging materials.

Proper sanitation in food processing and storage areas is the most effective weapon in the fight against rodents, since all packaging materials apart from metal and glass containers can be attacked by rats and mice.

Summary: Major causes of food deterioration include the following:

a. Growth and activities of micro-organisms, principally bacteria, yeasts and moulds;
b. Activities of natural food enzymes;
c. Insects, parasites and rodents;
d. temperature, both heat and cold;
e. Moisture and dryness;
f. Air and in particular oxygen;
g. Light;
h. Time.

Extrinsic factors controlling the rate of food Deterioration reactions are mainly:

a. Effect of temperature;
b. Effect of water activity (aw);
c. Effect of gas atmosphere;
d. Effect of light.

Fruit and Vegetable Preservation

Fresh Storage

Fresh Fruit and Vegetable Storage: Once fruit is harvested, any natural resistance to the action of spoiling micro-organisms is lost. Changes in enzymatic systems of the fruit also occur on harvest which may also accelerate the activity of spoilage organisms.

Means that are commonly used to prevent spoilage of fruits must include:

- care to prevent cutting or bruising of the fruit during picking or handling;
- refrigeration to minimise growth of micro-organisms and reduce enzyme activity;
- packaging or storage to control respiration rate and ripening;
- use of preservatives to kill micro-organisms on the fruit.

A principal economic loss occurring during transportation and/or storage of produce such as fresh fruit is the degradation which occurs between the field and the ultimate destination due to the effect of respiration. Methods to reduce such degradation are as follows:

- refrigerate the produce to reduce the rate of respiration;

- vacuum cooling;
- reduce the oxygen content of the environment in which the produce is kept to a value not above 5% of the atmosphere but above the value at which anaerobic respiration would begin. When the oxygen concentration is reduced within 60 minutes the deterioration is in practice negligible.

The following is a summary of some recent developments in Post Harvest technology of fresh fruits and vegetables (Source: Thompson, 1989).

Harvest Maturity: This is particularly important with fruit for export. One recent innovation is the measurement of resonant frequency of the fruit which should enable the grading out of over mature and under-mature fruit before they are packed for export.

Harvest Method: Considerable research is continuing on mechanical harvesting of perishable crops with a view to minimising damage. In fruit trees, controlling their height by use of dwarfing rootstocks, pruning and growth regulating chemicals will lead to easier, cheaper more accurate harvesting.

Handling Systems: Field packing of various vegetables for export has been carried out for many years. In the last decade or so this has been applied, in selected cases, to a few tropical fruit types. Where this system can be practiced it has considerable economic advantages in saving the cost of building, labour and equipment and can result in lower levels of damage into crops.

Pre-cooling: Little innovation has occurred in crop pre-cooling over the last decade. However high velocity, high humidity forced air systems have continued to be developed and refined. These are suitable for all types of produce and are relatively simple to build and operate and, while not providing the speed of cooling of a vacuum or hydrocooler, have the flexibility to be used with almost all crops.

Chemicals: There is a very strong health lobby whose objective is to reduce the use of chemicals in agriculture and particularly during the Post Harvest period. Every year sees the prohibition of the use of commonly used Post Harvest chemicals. New ways need to be developed to control Post Harvest diseases, pest and sprouting.

Coatings: Slowing down the metabolism of fruit and vegetables by coating them with a material which affects their gaseous exchange is being tested and used commercially on a number of products.

Controlled Environment Transport: Recent innovations in this technique have produced great progress as a result of the development and miniaturisation of equipment to measure carbon dioxide and oxygen. Several companies now offer containers where the levels of these two gases can be controlled very precisely.

8

Grape Production Systems

Pruning is one of the most important and most neglected practices in home plantings of grapes. Grapes need some form of support, and pruning (training) is necessary to develop the plant and to maintain it on the support provided. Regular, purposeful pruning is essential for controlling the number, position and vigour of fruiting canes and the yield and quality of the fruit. Grapes should be pruned during the dormant season, late November to March. Since winter injury of fruiting canes will occur to some extent, late winter pruning generally is preferred. If pruning is delayed until near bud swell, the cuts commonly ooze sap abundantly. Though not desirable, "bleeding" seems to be of minor importance.

Pruning at Planting Time

The best cane usually is selected and shortened to two buds. All other canes are removed. Sometimes a plant is especially vigorous and has one cane that will reach the lower wire or beyond. In this case, the cane may be cut at the height of the lower wire and tied tautly to it. When new shoots are about 1 inch long, remove them except for the upper two or three near the wire. From among the shoots allowed to develop, one will be selected to complete the trunk.

Second Year

Select the most vigorous cane and remove all others. If the cane is long enough to reach the top wire, cut it off at that height and tie tautly to the wire. A shorter cane would simply be pruned and tied to the lower wire. Plants having no canes long enough to reach the lower wire should be pruned as a newly set plant.

Early in the second growing season, some cane selection for the next year may be possible on the larger, more vigorous plants. When new shoots are 1 to 3 inches long, remove all but three or four arising from near each wire. Flower or fruit clusters are best removed as noticed. Plants bearing no fruit make stronger vine growth.

Third Year

In the dormant period preceding the third year, the more vigorous plants should consist of a main stem or trunk reaching to one or both trellis wires and having several canes. Considering position and vigour, select two of the best canes at each wire and remove the others. Shorten the selected canes leaving two to four buds on each.

The two to four buds left on the short canes will give rise to shoots that will bear some fruit. They also will be the source from which to choose next year's fruiting canes.

Fourth Year

Beginning in the fourth season, the plant should have two or more canes extending in both directions at each wire. Select a fruiting cane and where possible a renewal spur at each of the four arms. A typical plant before and after pruning. Note that renewal spurs were chosen from canes nearest the trunk. This is important for maintaining fruiting wood close to the trunk and keeping the whole plant within its allotted space. Where this is not practiced, the fruiting wood develops farther from the plant each year, crowding adjacent plants and weakening growth.

Select the fruiting canes with the following considerations in mind: moderate vigour (diameter and length), reasonably straight, originating near the trunk, and reasonably close to the appropriate trellis wires. All other growth is removed, including any sucker growth arising on the lower trunk. Shorten the selected fruiting canes leaving six to 10 buds. Leave more or fewer buds depending on plant vigour – 10 to 12 buds on vigorous plants, four to six on weaker plants.

In the following years, pruning will not be greatly different from fourth year pruning. Where a renewal spur was present during the preceding growing season, it is frequently the nearest source from which to select a fruiting cane. Thus, the old arm with numerous canes attached can be removed with one cut made near the trunk. If no renewal spur is present, select a cane elsewhere and if possible, leave a spur for use the following year.

The number of buds left per cane may be increased to about 10 to 12 on plants of good vigour. Again, number of buds left should be adjusted according to plant vigour.

Four-cane Kniffin system

The four-cane Kniffin system is similar to an ordinary two-wire fence. Wooden end posts should be at least 6 inches in diameter and from 8 to 8-1/2 feet long. When set about 3 feet deep and properly braced, the wires can be stretched fairly tight. Other posts may be smaller and need not be set as deeply as end posts. Oak posts treated with a preservative are satisfactory. Steel fence posts may be used except for the end.

Umbrella Kniffin System

This system provides good conditions for producing high quality fruit. The fruit and foliage is held well up on the trellis for good sunlight exposure and air circulation. These conditions aid in reducing diseases. The long fruiting canes used make the system best adapted to vigorous varieties or in situations conducive to vigorous growth.

Early training follows the procedure described previously, and the same two-wire trellis is used.

Develop the trunk to a point near the top trellis wire, removing canes that arise near the lower wire or below. Under average conditions this will require two years, but some plants may reach the top wire in one. In the following growing season, shoots will develop from buds along the upper trunk. The growth of these shoots can be improved by early removal of those which develop below.

In the next dormant season, select two canes (one on each side of the plant) which originate about midway between the two wires. Remove all others. If the plant has made vigorous growth, the canes can be left 10 to 12 buds long.

The length should permit bending the cane over the top wire and then down and outward from the trunk to the lower wire. Tie the cane snugly to the lower wire. If the canes are short and small in diameter (weak), shorten the two selected canes to spurs with two to four buds each.

Following this pruning, a moderate crop of fruit and several strong canes should be produced. The most vigorous canes will develop from buds between the trunk and the top wire. This provides good fruiting wood in the desired position for the next year. The remainder of the fruiting cane, being inverted, tends to make less growth and be more fruitful.

Each year thereafter, select three or four of the best canes, cut back two others near the trunk for renewal spurs and remove all others. Bend each fruiting cane individually over the top wire and extend to the lower wire and tie firmly. Cut off any portion of the fruiting cane that extends below the lower wire. The fruiting canes should be spaced along the trellis and not bunched together.

Cordon Systems

What is being described as cordon system training here goes by a variety of names, including single curtain cordon, bilateral cordon and high cordon, but the basic principles among these specific systems remain constant. This system probably is the one to choose for most Missouri conditions for the following reasons: it allows for excellent light and spray penetration, yields high-quality fruit, is easy to prune and pick, requires little or no tying and provides some frost protection with high wire. On the other hand, the cordon systems are not suitable for winter-tender cultivars, must be combed, can suffer from wind damage, exposes the fruit more to birds and sun, and some cultivars just are not adaptable to it. These systems can work well for homeowners who are having trouble managing the vines on the lower wire in a four-arm Kniffen.

While the trunk is the only permanent above-ground part of the grape plant in the Kniffen systems, the trunk and cordons along the top wire are permanent in the cordon systems. In a bilateral cordon, the plant is trained in both directions along the wire from the trunk, while in the single cordon system, the plant is trained in a single direction. In addition, the cordon systems require only one high wire, although some growers will use a lower wire to help train new vines upward toward the high wire.

The plants are treated similarly to the four-arm Kniffen in years one through three. However, in the third year, the canes you select will become the permanent cordons. Pruning in years beyond the third year involves selecting spurs that will fruit next year, shortening them and removing all others. In general, on a 3-year-old plant, four to 10 fruiting spurs are left on each cordon, each with one or two buds. As the plants age, six to 12 fruiting spurs with two to five buds on each spur are selected to remain on each cordon. Ideally, these fruiting spurs are spaced 6 inches to 1 foot apart. As with the Kniffen systems, the number of fruiting spurs and buds depends on the vigour of the variety grown; more may be left on vigorous plants and less on weaker plants.

In cordon systems, the permanent cordons should be periodically renewed, especially if a cordon has been severely injured by low temperatures, insects or diseases. Simply train another well-positioned shoot near where the old cordon arises, grow it for a year, then remove the old cordon the following year.

"Combing" should be done when using the cordon systems. In combing, the shoots that develop from the retained buds are vertically positioned (combed) downward to prevent shoots from growing over one another. If a lower wire is in place, these combed shoots often are tied to the lower wire.

Fan System

This system is useful for training to walls and fences, although a special trellis or arbor could be used. A plant pruned and trained to this system has several upright canes branching from arms on a very short trunk. Since the upright training promotes good growth, weak plants (due to variety of soil conditions) would be well adapted to this system.

Many of the developing shoots droop to the ground forming a more effective screen. A disadvantage is that the fruit developing near the ground will become dirty or possibly decayed.

The newly set plant is pruned to a short two-bud spur. If the plant grows well in the first season, both of the canes (one from each bud) may be left after shortening each to three or four buds. A weaker plant may be left with two spurs of two buds each. At the beginning of the third season, the plant may have four or more good canes. Select three or four of the best for fruiting canes. If others are present in a fairly low position, leave two or three of them as two-bud spurs and remove any others. Shorten the fruiting canes to the height of the trellis and tie them at the tip. Position the canes in a spreading fanlike arrangement.

Some fruit will be produced in the third season

The number of fruiting canes may be increased to six or eight in subsequent years, depending on length of the canes (number of buds) and plant vigour. Select fruiting canes from renewal spurs where possible. Otherwise, select them from near the base of last year's fruiting cane and if possible leave three or four spurs for use the following year.

Arbors

Many prefer to grow grapes on an arbor, thus combining the benefits of fruit production, shade and ornamental effects. Vines grown on arbors generally produce less low-quality fruit than traditional systems do. Suitable arbors can be in an almost endless number of designs. A more important consideration is that the arbor be constructed of durable materials requiring minimal maintenance.

In general, plants are placed on both sides of an arch-like structure and trained to grow up and over to about mid-point of the top. Providing this amount of foliage cover re-quires a larger and taller plant than is necessary for an ordinary trellis. In this situation there is a tendency to prune too lightly, if at all. Often the end result is a tangled mass of multiple trunks, numerous canes, weak growth and poor fruit production. Training the plants to a single trunk and leaving relatively short horizontal fruiting canes is a suitable method for most situations.

Develop a portion of the trunk each year by tying an uppermost vigorous cane in a vertical position. At the same time, select fruiting canes at intervals of 2 to 3 feet. These should be limited to five or six buds to favour development of the upper trunk and canes. The selection and use of renewal spurs also is valuable for maintaining a source of fruiting wood close to the trunk.

Single-trunk us. Two-trunk Training

Commercial grape growers often will train two parallel trunks in the systems that have just been described. In essence, each trunk makes up for half the canopy. For instance, in the case of the four-arm Kniffen, one trunk might produce the canes for the right side lower wire and the left side upper wire, and vice versa for the other trunk. Growers do this as a hedge against stress, particularly winter damage; if one trunk is killed, production from that plant is not completely lost. Two trunks also allows for renewing a trunk without losing production from the plant.

Tying

Use a material that will last the entire season. Binder twine or a lighter jute twine is satisfactory. Plastic covered wire ties used for plastic bags, etc. also could be used.

Tying should be done during relatively mild temperatures, but before buds have swollen or started growth. Canes that are very cold or frozen are brittle and easily broken. Such breakage is particularly likely when sharply bending the canes as required with the Umbrella Kniffin system. Buds that are quite swollen are easily broken off.

With the Four Cane Kniffin system, wrap the individual canes around the wire once or twice, depending on cane length, and tie just short of the last bud. Two wraps of twine with a firm double knot will hold a cane securely.

Terms

Some knowledge of grapevine parts is helpful in understanding pruning details.

- Trunk: This is the main permanent stem of the plant.
- Shoots: Immature, soft stem growths of the current growing season are shoots. Shoots arise from buds on wood one or more years old and bear leaves, flowers and fruit.
- Canes: These are the mature shoots, those which have become woody after growth has ceased. Fruiting cane merely refers to a 1-year-old cane that is capable and suitable for bearing fruit.
- Arms: These might be called temporary side extensions of the main stem. They are the basal portions of former canes that were left after pruning.
- Spurs or renewal spurs: These are 1-year-old canes (preferably originating near the trunk) shortened to two buds. Shoots (and later canes) develop from the spur buds. One of these is selected as a fruiting cane for the following season, thus "renewing" the fruiting wood.
- Sucker: This shoot (or cane) usually arises from the lower part of the trunk.

Pruning and Training Persimmons

Persimmon cultivars differ markedly in their vigour and growth habits. The major cultivar, Fuyu, is moderately vigorous and if left unpruned will reach about 5 m in height and spread in 10 years. Other cultivars such as Matsumoto-Wase-Fuyu, Izu and 20th Century are less vigorous but are open and spreading.

Training Systems

There are several reasons for training trees to a specific system:

* To allow light into the canopy to improve fruit colour and sugar levels.
* To simplify cultural operations such as pruning, thinning and harvesting.
* To reduce limb breakage and fruit rub.

There is little information available on the suitability of the different training systems for persimmon. We can, however, make some general comments based on limited overseas and local experience.

Open Vase

This system has been tried in Australia. Trees are very open and prone to limb breakage and sunburn.

Tatura trellis, Lincoln trellis and their variations: Trellised systems are desirable because they should produce the best quality fruit of all the training systems available. Also, they should help limit tree size so bird netting can be more easily erected. However, persimmons would probably outgrow these systems in our climate unless widely spaced. Limited overseas work suggests this may be the case. The development of growth retardants or dwarfing rootstocks would make these systems more feasible. These systems should only be experimental at this stage.

Palmette (Trellised or Untrellised)

This system should be suitable for trees spaced at about 4.5 m x 3 m. A trellised palmette would be advisable with wires at 50 cm, 100 cm, 150 cm and 200 cm high.

Free Standing Central Leader

This is a fairly inexpensive and simple system to develop. For these reasons it is the system most commonly used. Disadvantages are that trees become fairly tall and, because it is untrellised, limb breakage and fruit rub can be a problem. Spacing will vary with cultivar and soil depth. A suitable spacing for Fuyu on good deep soil is about 6 m x 5 m. Matsumoto-Wase-Fuyu, Century and Ichikikei Jiro could be spaced at about 5 m x 3 m.

Summary

At this stage of the industry's development, it is suggested that growers adopt a low-cost approach to training systems. Either the palmette or free standing central leader are practical options. The trellised palmette is probably the best of the practical alternatives. Other trellised systems such as the Lincoln and Tatura need to be tested.

Flowering, Fruiting and Growth Patterns

Before you can understand how to prune the persimmon you must understand the growth and fruiting habits of the tree.

Flowers are produced on new growth in the spring period. This new growth emerges from one-year-old wood. We can refer to this one-year-old wood as fruiting wood. It is the light-brown growth you see on trees in the dormant period (winter).

The best fruiting wood is about 1 cm thick and 30 cm long. It usually has about ten buds. The end four to six buds will produce new growth in September to October, and these will have flowers. This is what we refer to as a 'tip-bearing habit'. The bottom two to three buds will produce weak non-bearing laterals. These weak laterals usually die out.

The flowers will be produced in the bottom (first) few leaf axils of the new growth. Many of the flowers will be shed or must be thinned out to leave one to two fruit per lateral.

Vegetative growth will continue until about December. You may get a second summer flush in vigorous or young trees. The spring growth which is carrying fruit will produce flowers towards the tips the following season.

If no pruning is done the fruiting wood will be produced further and further away from the main leader. There will be more but smaller fruit. Because fruit is produced on the ends of fruiting arms, limb breakage and fruit damage would be more likely.

Pruning Young Trees

The aim of pruning in the first three to four years is to develop the main framework of the tree.

A grafted persimmon tree, as obtained from a nursery, usually consists of a single stem. Pruning is needed to encourage a strong central leader with tiers of branches at about 70 cm intervals up to the trunk. The usual times to prune are August and late November.

The first pruning is just after planting. If there is no branching within 50 cm of the ground, cut the single stem at that height. If the central leader is not the highest point, the tree should be pruned to ensure it is.

Over the next two to three years the central leader should be pruned about ever 70 cm to encourage branching at those positions. Long laterals should also be tipped back. Prune out strong growths that are not needed. Always ensure the basic shape of the tree is developed without branches crowding one another. Avoid severe pruning as it only induces excessive growth at the expense of fruiting.

Pruning Bearing Trees

The main aim of pruning a bearing tree is to prevent the development of long drooping leaders. It also helps to keep the tree open and manageable. Pruning is normally done once per year, in late August. The procedure is as follows:

* Prune off low shoots and suckers.
* Break of any dead wood within the canopy.
* Prune out strong growths not needed.
* Prune back long laterals to about 70 cm or so.
* Prune out some laterals if too crowded.
* Prune some fruiting arms back to one or two laterals closer to the main leader. Don't do this to all fruiting arms but rather keep a balance of old and new fruiting arms. Replacement of fruiting arms shouldn't be needed until about year 4 or 5, but every year from then onwards.

Trees may need to be lightly tipped in late November if they are vigorous. If the trees are vigorous, the fertiliser programme should also be adjusted to reduce vigour. Because persimmons have a tip-bearing habit you should be careful not to tip fruiting laterals back too much. Pruning cuts over about 20 mm in diameter should be treated with a white water-based paint containing a mould inhibitor.

Crop Thinning

Trees can go into a biennial bearing habit if allowed to crop very heavily in any one year. Fruit size will also be reduced if trees set heavy crops. Fuyu requires about 15 leaves per fruit to give adequate size. Insect damage is increased where fruit forms tight clusters.

To avoid these problems, remove a part of the crop in October, just after fruit set. Remove the fruit closest to the base of the shoot as these are more likely to be shed. Only one fruit should be left on most laterals. Laterals longer than about 200 mm could carry two or three fruit as long as they are well-spaced.

Environment and Horticultural Plants

Horticultural plants grow in response to their environment. Plants are cultivated from many areas of the world. Each has a different set of environmental conditions to which the plant has become adapted. Much of horticulture is concerned with modifying plant environments to enhance growth. This chapter covers several of the major environmental factors that influence plant growth and the influence of climatic variations on horticultural plants.

Climate

Climate is the combined effects of temperature, light and elements of moisture such as clouds, rain, hail, snow and wind. Climate usually refers to long term weather patterns in a region or as it is often called, macroclimate.

Temperatures are greatest at the earth's surface near the equator where the sun's rays strike directly. As you move north or south from the equator it gradually becomes cooler. Altitude also influences temperatures with cooler temperatures occurring at higher altitudes. Climatic patterns are also altered by large bodies of water nearby. More energy is required to raise the temperature of water than air. Water releases this energy as it cools. Thus, water acts as a buffer for nearby land areas and reduces temperature extremes. Inland areas of the U.S. lack the protection of large bodies of water and experience high summer temperatures and cold, arctic fronts in the winter.

Mountains also alter climate by blocking natural wind and moisture movement. Several areas of the Pacific northwest have wet, rainy areas on the ocean side of mountains and dry areas on the inland side. Vegetation also influences climate. Forested areas have a higher relative humidity. This and additional evaporation of water from tree leaves has a cooling effect.

Human activities also influence climate. Large reservoirs have been constructed in many areas. Metropolitan areas are warmed by combustion used in heating and by buildings, streets and other structures absorbing and holding heat. Dust, environmental pollutants and smoke also alter the climate. A concern in many areas of the eastern U.S., Canada and Europe is acid rain, formed as rainfall absorbs acidic materials from air pollution. All of these macroclimate differences have an effect on plant species growing in an area and may cause climatic variations.

Microclimate refers to the climate in a small area such as near a plant or in a yard or garden. Cold air is heavier than warm air and tends to settle in low areas creating a temperature inversion. Temperatures in low areas can be several degrees cooler than surrounding hillsides. These pockets of cold air are called frost pockets. Orchards are generally located on hillsides to avoid areas likely to damage flowers in early spring. Selecting a home site on a sloping area can also avoid low temperature injury to ornamentals in the fall and spring. Carefully locating plants around a home can take advantage of microclimate differences. Shade loving plants can be grown on north or east exposures and plants can be protected from the blast of west or north winds by planting them on the protected side of the house. Horticulturists need to be aware of the influence of climate in a particular area, both macroclimates and microclimates. Selection and growth of horticultural plants is directly influenced by climate.

Temperature

Temperature largely determines what can be grown in a particular area. Photosynthesis (the conversion of solar energy for use by the plant) and respiration (the using of stored energy compounds in the plant for growth and development) are regulated by temperature. With most biological reactions each 10°C (18°F) increase in temperature doubles the rate of the reaction. Therefore, we expect the rate of growth at 68°F to be twice as fast as at 50°F. This relationship is known as Q10. Each plant has an optimum temperature for growth and development. There is a maximum temperature where plant growth stops and permanent injury to the plant occurs. Likewise, there is a minimum temperature where plant growth stops and freezing or chilling injury occurs.

Cool Season Plants

Depending on their area of native adaptation, plants prefer cooler or warmer temperatures. Cool loving plants tolerate temperatures slightly below the 32°F. Growth is poor at warmer temperatures. Examples of cool season plants include vegetables (cabbage, broccoli, lettuce, radishes, peas), flowers (crocus, daffodil, tulips, violets), fruits (apples, pears, plums, American grapes) and shrubs (forsythia, lilac, spirea, honeysuckle).

Cool season plants prefer average daily temperatures below 70°F. Premature seed stalk formation or bolting of biennial vegetables is a problem for cool season crops. Cool temperatures following planting compresses their natural growth cycle and seed stalks often form later in the season.

Warm Season Plants

Some plants prefer temperatures above 70°F and are usually injured by freezing temperatures. Sometimes these plants are called nonhardy or tender plants, while those that can withstand freezing temperatures are called hardy plants. Examples of warm season crops are vegetables (beans, tomatoes, sweet corn, melons), flowers (roses, lilies tropical foliage plants) and fruits (peaches, apricots, sweet cherries).

Freezing Injury

When the temperature drops below 32°F there is a chance of injury to some plants. This is frequently called frost injury. Frost is ice formation on a surface while plant injury occurs only when ice forms within the plant. Generally, the period just before dawn is the time when the temperature reaches the lowest point and freezing injury occurs. Freezing injury results in an immediate water soaked, blackened appearance that quickly turns brown and dies. It is a common horticultural practice to cover plants with insulating materials or sprinkle plants with water to prevent freezing temperatures from occurring within the plant. While water from a sprinkler may feel very cold it has some heat energy that is passed on to the plant. There is a lot of current research on treatments to allow plants to be exposed to lower temperatures without injury and to understand the mechanism of freezing injury to plant tissues.

Chilling Injury

Injury can occur to warm season plants or products exposed to low temperatures. This low temperature injury is called chilling injury and results from a malfunction in the normal plant growth processes rather than the freezing of water within the plant. Crops such as cucumbers, tomatoes, tropical fruits and most tropical foliage plants, are subject to chilling injury. The development of chilling injury usually results in rapid respiration, the development of rots and molds, bitter or "off flavours" and abnormal colours. Chilling injury may occur at temperatures below 45-50°F in many plants or products. You see it when a banana turns black after being placed in a refrigerator.

Rest Period

There are periods when a plant or seed does not grow despite favourable conditions. These periods are called rest periods. They usually coincide with the natural period of dormancy for perennial plants during the winter. During this time deciduous plants lose their leaves and herbaceous plants are frozen back to the ground. Buds of woody plants do not grow due to physiological and biochemical processes. Following a period of cold, the buds break (end) their rest period and continue growth. The temperature and hours needed to break dormancy has been determined for many plants. Fruit varieties are grouped by their requirement of hours at or below a critical temperature (usually 40°F) to break the rest period. Desirable environmental conditions are necessary to resume growth following the breaking of the rest period. Many temperate plants produce seeds with a natural rest period following ripening. This rest period can only be broken by exposure to cold temperatures for a period of time. This process is called stratification. We can artificially stratify seeds by storing them in a moist medium such as peat moss or sand at near freezing temperatures (usually 40°F) for a number of weeks. This breaks the normal rest period and allows the seeds to germinate uniformly.

Woody Plant Survival

An example of hardiness has already been provided for annual plants tolerant to low temperatures. Woody plants can also tolerate low temperatures to varying degrees. Native plants have, through the years, adapted to withstand low temperatures and are said to be hardy to the area where they evolved. The level of low winter temperatures that a particular type of plant can withstand has been determined for most woody plant species.

The United States is divided into areas called plant hardiness zones. The zones are numbered 1-10 from colder to warmer areas. This map was developed by the US Department of Agriculture. Woody ornamental plants survive average normal low winter temperatures in these areas and plants can be characterized by their adaptation to these climatic zones. It must be remembered that these zones are based on the average of all of the lowest winter temperatures recorded over a 50 year period. In certain extreme years, such as in January 1994, severe damage to plant material occurred because the temperatures were so much colder than normally would be expected. Damage also frequently occurs to normally hardy plants when an unusually early frost or freeze hits plants before they are fully dormant or in the spring after they have begun to lose dormancy.

Modifying Temperatures

Much of horticulture is involved in altering the environment to improve plant growth. Areas of horticultural production have developed in areas where natural protection from temperature extremes exists. Such protection is provided by large bodies of water such as the western area of Michigan and the western areas of New York bordering the Great Lakes. Buildings, streets, etc. in large cities absorb heat and reradiate it at night providing several degrees of warmth against the first freezing temperatures of the fall. Growing plants in a managed environment like a greenhouse was developed to provide a favourable temperature during adverse periods. There are other examples of treatments horticulturists use to modify temperature. These including the use of mulches. Black or clear plastic absorbs heat and transfers it to the soil encouraging earlier plant growth in the spring. Hotcaps or hot-tents-paper or plastic covers also serve as miniature greenhouses to modify the microclimate around transplants in the spring.

Sheets of fiberglass or porous plastic are frequently placed over rows of plants to warm them in the spring. This type of structure is called a cloche (pronounced klosh) or row cover. Another example or a specialized plant growing structure is a hotbed or coldframe used to start seedling plants. Water mist is frequently used as a means of reducing freezing temperature injury. The protection provided by water mist is based on the principle that as water freezes it gives off heat. Thus, as the water freezes it gives off enough heat to protect plants for a few degrees. Other examples of frost protection include smudge pots, stoves and air turbulators used in orchards to force warmer air around plant surfaces. Horticulturists continue to study ways to understand the mechanisms of temperature injury, ways of improving the adaptation of plants to temperature extremes and ways of altering the temperature around horticultural plants to improve their growth.

Water

A brief drive through the desert, failure to water a pot of flowers, or wilted lettuce on a salad bar reminds us of the importance of water for plant growth and development. Just as with temperature extremes, plants have adapted to grow in areas of low water availability. Thick, fleshy leaves, layers of wax, specialized structures for water storage and other mechanisms help plants survive in areas of limited water. Such plants are called xerophitic.

Water in the Plant and Soil

Water constitutes an important part of plants. Living plants need a constant flow of water from the roots to leaves to continue their life processes. Water in plant cells provides pressure on cell walls, called turgor pressure, keeping the plant rigid. Water is an important constituent of photosynthesis. Water flows from the root, up the stem and into the leaves in specialized conducting cells called xylem cells. Movement of nutrients and other constituents flow in water in the plant as well. Pressure from water in the cell enables cells to enlarge and expand. Water not used by plant cells evaporates and moves out of the leaf through small pores or openings in the leaf surface called stomata. Evaporation of water from the leaf is called transpiration. Evaporation of water from the leaf has a cooling effect and reduces temperatures in the surrounding environment.

Water moves into root cells from the soil. Soils differ in the amount of water that they can hold. Sandy soils hold less water than loam or clay soils. Water available to a plant from a particular soil is called the soil water potential. Some water is held so tightly by the soil that the plant cannot extract it. Water that the plant can extract from the soil is called available water. The root system develops in the soil to obtain water for the plant. Sudden flooding of the soil cuts off oxygen to the root system and causes root death. Plants in flooded locations can die quickly from such injury. Too much water can be as serious a problem as not enough water. For this reason, a greenhouse manager must carefully monitor the water level in containers to provide the proper amount of water to meet the crop's needs. In field situations, drainage tiles or pipes are often installed to allow excess water to be removed. Water is lost from soils by evaporation from the soil surface. Evaporation of water from soil combined with loss of water through the plant (transpiration) is referred to by the combined word evapotranspiration, or loss of water by plant and soil combined.

Applying Water to Plants

Horticultural production has developed where water is abundant. Applying water to supplement rainfall is a standard practice for many horticultural crops. Irrigation should replace water removed by plants. Considerations in irrigation amounts and frequency depend on the needs of the crop, the depth of roots and the ability of the soil to absorb water. A certain portion of the water is lost to evaporation before it reaches the soil. On a windy day a sprinkler system putting out a fine mist may lose 25-30% of the water applied before it reaches the soil. Drip or trickle irrigation allows small amounts of water to be applied frequently to a portion of the root zone. This reduces the amount of water needed to grow a crop.

Effects of Limited Water Availability

For many horticultural plants there is a period of critical water need when water may not be available. For most plants this is during the period of flowering and fruit development. A symptom of water need is wilting of the foliage. Plants are affected by stress before the external symptom of wilting occurs. Emphasis on water levels necessary to support plant growth and scheduling irrigation to minimize water use is important. Water needs can be predicted from climatic factors such as temperature, humidity, wind and solar radiation combined with particular characteristics of the crop and soil type. Microcomputers assemble various weather factors and develop predictions and schedules for adding supplemental water. Fluctuations in water available can cause sudden splitting of cherries, tomatoes and cabbage. Uniform water supplies are necessary to prevent these problems.

Light

Light is essential for photosynthesis. Green plants are the original solar collectors. They transfer energy from the sun to a usable form of energy for the plant. Light varies in intensity, quality and duration or photoperiod.

Intensity

The intensity or amount of light available has a direct influence on the amount of photo-synthesis that can occur. Some plants have adapted to growing in areas of low light availability such as a jungle floor. This is why tropical foliage plants grow well in limited indoor light. Grasses are native to the open plains and require high light intensity for growth. When light intensity is lowered, the rate of photosynthesis is reduced. When the level of photosynthesis equals respiration, it is referred to as the compensation point. Below compensation point growth ceases.

Sudden exposure to high light intensity can cause leaf scorch or sunburning. The symptoms are large, brown, dead areas on leaves. Plants growing in low light have a thinner layer of wax on leaf surfaces. This helps the leaf to capture more light. However, moving it suddenly into bright light results in rapid water loss. High light intensity can also cause fading of flowers, especially in hot summer weather. An area of special interest to the greenhouse industry is the use of high intensity lights to supplement natural light and to extend the photoperiod. Crops of high value make the expense of installing and operating high intensity supplemental lights economical.

Light Quality

Light quality refers to the "colour" of the light or the portion of the light spectrum. The biochemical constituents of chlorophyll, the compound responsible for photosynthesis, absorb only light from particular portions of the light spectrum. Artificial lights supply light of particular portions of the spectrum. Light absorbed by plants is from the blue-violet and orange-red parts of the spectrum. Specialized lamps have been developed to provide light rich in wavelengths from this part of the spectrum. These lamps have the additional benefit of creating dark green foliage and a deep, intense flower colour. Excellent plant growth can be achieved with ordinary fluorescent bulbs. Other portions of the spectrum such as ultraviolet light are important in the coloration of some fruits and the development of autumn colour in leaves.

Relative Humidity and Wind

Relative humidity is the amount of water vapour in the air. It is closely related to temperature. As the temperature goes down, relative humidity goes up. Relative humidity is measured as a percent. When it reaches 100% we have rain, snow, fog, or sleet. This is why it rains as a cold front goes through.

As the relative humidity drops, evaporation increases. Plants leaves give off more water at 40% than 80% relative humidity. This water loss increases if the wind is blowing. Tropical foliage plants are native of humid, jungle environments and often suffer in the dry, indoor air in winter. Grouping plants together and placing plants on a layer of pebbles in standing water helps increase relative humidity. Lowering the interior temperature of homes is beneficial in reducing some of the low humidity problems with plants.

Humidity is often related to the development of certain plant diseases. Warm, damp conditions are ideal to encourage the growth of fungi. Many fungal spores must have water to begin growing. Damping off is a common problem for seedlings. It results from fungi rotting small plants at the soil line. Increasing air circulation to lower humidity and reducing watering can generally stop the problem.

Wind is a problem with plant production in some areas. In addition to blowing away valuable soil, wind can also cause injury to tender plant stems by sandblasting the stems at the soil line. New transplants and newly planted trees are often shaded as a protection from hot sun and wind. In many areas of the Great Plains of the U.S., the planting of rows of trees in a windbreak protects homes and animals. Strong winds in storms damage many trees and shrubs each year. Wind damage is especially severe when it is in combination with ice. The plant species is also important. Trees such as silver maple, Siberian elm and willow are more frequently injured by wind than oak, walnut and sugar maple.

Interrelationships

Plants are influenced simultaneously by all segments of their environment. As light intensity increases, temperature generally increases with a corresponding increase in photosynthesis and water loss by the plant. Many times horticulturists must alter one environmental factor as another changes. When light levels are reduced, a greenhouse manager will often lower temperatures. Without the temperature reduction, plants become tall and spindly. The reduction in temperature limits the problem. Although a lot is known about environment factors that influence horticultural plants, much is still to be learned. Urban conditions influence plants in both positive and negative ways. Learning to compensate for these negative influences will become more important in the future.

9

Strawberry Cultivars and their Culture

No fruit is more likely to provide home gardeners with success and satisfaction than strawberries. Strawberries are well adapted to our climate, require a very small investment, and will produce a good crop of fruit about 13 months after planting. Strawberry plants will bear every year if blossoms are protected from frost by mulching and will produce a good crop of berries for up to four years.

Before Planting

Select the Best Site

Strawberries need full sun. Soil should have good surface and internal drainage. Ideally, the organic matter content should be about 5 percent, although strawberries will grow in soil that has as little as 2 percent organic matter.

Take a Soil Test

During the fall before planting, take a soil sample to your local MU Extension centre and have it tested. If the soil test results indicate that additional nutrients are needed before planting, incorporate them in late winter. If the organic matter content is low, leaf mold, compost, rotted manure or aged sawdust can be incorporated to improve the soil tilth.

Decide how many to Plant

You can expect to harvest about 1 quart of berries from each plant you set out plus its daughter plants if grown in a matted row system. Twenty-five to 50 plants will provide plenty of fresh berries for a family of four. For freezing berries, set out an additional 50 to 100 plants.

Purchase well-adapted Cultivars

Many strawberry cultivars are suitable for growing under Missouri conditions. In selecting varieties consider.

- Disease resistance
- Dessert and freezing quality
- Yield
- Season of ripening.

Always purchase new plants rather than getting plants from neighbors or relatives. New plants are certified to be free of disease and should be free of viruses, whereas those obtained from a neighbor may harbour disease and insect pests. Buy your plants from a nursery with a good reputation. Consult your neighbors to see what their experience has been with various nurseries. If you are the first in your area to order strawberries, look at catalogues from a variety of nurseries. Do they guarantee disease-free plants? What is their policy if you get poor-quality plants from them? Do their descriptions give cultivar limitations as well as positive traits? Do their prices seem fair? Within reason, it is always better to pay a little more for high-quality plants than to spend less for cheaper, weaker plants.

Popular and better Adapted Varieties for Missouri

This list is not intended to include all varieties that will grow in Missouri.

- Earliglow: An early-season cultivar that produces berries that have outstanding flavour. Fruit is attractive with high quality, and it freezes very well. The primary berries are medium-sized and secondary berries are small. The plant is vigorous and usually winter hardy. Shows resistance to leaf and root diseases.
- Cardinal: Vigorous plants produce high yields of large fruit in early midseason. Fruit are firm and have good flavour. Plants are winter tender but often are planted as far north as central Missouri. Plants are not resistant to red stele but are resistant to leaf diseases.
- Honeoye: Berries are medium-large and bright red but sometimes lack good flavour. High yields are harvested in early midseason. Tolerant of most foliage diseases but not resistant to red stele or verticillium wilt. Excellent for freezing. Winter hardy.
- Redchief: Berries are medium-large, sweet and have exceptional quality. This cultivar is resistant to five strains of red stele root disease and verticillium wilt. It produces runners freely, and the berries are harvested in early midseason. Excellent for freezing.
- Surecrop: Berries are large, light-coloured and tart with fair quality. This is one of the most productive, vigorous and disease-resistant cultivar available. Berries ripen in early midseason.
- Allstar: Berries ripen in late midseason. They are large, glossy, orange-red, firm and very good quality for fresh eating. Allstar is resistant to red stele and verticillium wilt and is tolerant of leaf diseases.

- Guardian: A late midseason cultivar with moderate to high yield potential of medium to large fruit. Excellent fruit firmness and very good flavour. Susceptible to leaf spot but resistant to leaf scorch, red stele and verticillium wilt.
- Jewel: Plants produce high yields of large wedge-shaped fruit with good firmness and very good flavour that ripens in late midseason. Plants should be hardy in most Missouri winters, but runners do not grow excessively. It has some resistance to leaf diseases but no resistance to red stele or verticillium wilt.
- Lateglow: A late-season cultivar with high yield potential. Fruit are very large with excellent firmness and very good flavour. The plants are resistant to red stele and will tolerate most Missouri winters.
- Sparkle: An old favourite with fruit that ripens late in the season. Berries tend to be soft and medium-sized, but have very good flavour. Plants bloom late so they may escape spring frost injury. They are resistant to the most common strain of red stele but susceptible to leaf diseases and verticillium wilt.

Everbearing and Day-neutral Cultivars

Traditional everbearing cultivars such as Ogallala and Ozark Beauty are not truly everbearing. They tend to produce a crop in the spring and a small crop in the fall, with little or no crop in between. Ogallala berries are medium sized and dark coloured with a fair flavour. Ozark Beauty produces medium-sized, light-coloured berries of average quality and has been extremely variable in its performance in Missouri.

Day-neutral cultivars can potentially produce flower buds, and fruit can be harvested about every six weeks regardless of day length. However, these cultivars do not initiate floral buds when the temperature exceeds 85 degrees Fahrenheit. Thus, day-neutral cultivars such as Tribute and Tristar generally produce a spring and fall crop in most of Missouri. Tribute plants produce moderate to high yields. Berries are large, have excellent firmness and very good flavour. The plants have good winter hardiness, are resistant to red stele and verticillium wilt and tolerant of the leaf diseases. Tristar plants produce moderate yields. Berries are medium to large and have excellent firmness and outstanding flavour. These plants have good winter hardiness, are resistant to red stele and verticillium wilt and are tolerant of the leaf diseases.

At Planting

Plant as early as soil is workable in the spring – early March to early April. Test for workability by digging a shovel full of soil and tightly squeezing a handful. Does the ball of soil break easily? If so, it is ready to till. Add 10 pounds 5-10-5 or 8-24-

8 fertilizer per 1,000 square feet and work into the upper 3 to 4 inches of soil. Plants can now be transplanted. Space rows 42 to 48 inches apart; plants in the row should be spaced 30 to 36 inches apart.

When Setting out Plants

- Keep the roots moist at all times
- Spread roots out in hole
- Plant at proper depth
- Water plants (you can add 1 tablespoon of a water-soluble fertilizer per gallon of transplant water).

First-year Care

Remove Blossoms

Remove blooms on newly planted strawberries. If berries are allowed to develop, they deplete the food reserves of the plant and delay runner production.

Cultivate

Perhaps the most important practice in a new strawberry planting is cultivation. For the first six weeks or so it should be frequent – perhaps once a week. This does two things:

- Kills weeds
- Loosens soil for better runner penetration. If necessary, herbicides can be an effective tool in controlling weeds in strawberries.

Train Runners

To establish manageable rows, train runners to root in the row rather than the aisle between rows.

Irrigate

Strawberries can survive all but the most severe drought periods. However, irrigate to get the highest yields of large berries. Strawberries should have a minimum of 1 inch of water per week – more during prolonged hot, dry periods. A simple rain gauge will determine weekly rainfall. Irrigate to make up for rainfall deficiencies.

Mulching

Winter mulch is used to prevent alternate freezing and thawing of the soil and subsequent heaving of the plants out of the soil. Other benefits of mulching include help in controlling weeds and grass, protection from severe cold temperatures and help in keeping berries clean during harvest.

Kinds of Mulches

Many materials can be used for mulching; in Missouri the two most commonly used are straw and aged sawdust. Obtain straw shortly after wheat harvest. Loosen bales and soak with water. This and subsequent rains should germinate most of the grain before it is time to apply the straw.

When to Apply Mulches

Apply straw mulch in late November or early December – after a few days of temperatures down to 20 degrees Fahrenheit. Straw should be sifted loosely over the plants, just enough to cover them from view. After a week of settling, add additional straw where necessary. When sawdust is used for a mulch, try to maintain a 1-inch mulch depth. No additional winter protection is needed.

Second-and Third-year Care

Remove Mulch

Remove straw mulch in the spring as the weather begins to warm (before bloom). Check closely after each warm period; if the strawberry foliage begins to show yellowing, remove the mulch. Rake mulch toward row aisles, but leave a light sprinkling on the plants – they will push through. If spring frosts threaten, rake the mulch back over the plants. But be sure to rake off the mulch during the day.

Control Insects and Diseases

Many pest problems can be avoided by using good cultural practices as described above. When insects or diseases can no longer be controlled culturally, pesticides may be applied.

Harvest

Pick berries daily if possible and pick all berries that are ripe. Toss out all moldy berries. This will help prevent rots from spreading.

Renovate and Fertilize

After harvest decide what to do with your strawberry bed. The two options are:

- Simply re-fruit the bed by renovating the plants immediately after harvest. First, remove weeds from row middles. Next, mow old tops off of the strawberry plants to renovate or renew the bed. Set lawn mower high enough to remove strawberry leaves but not injure crowns. Then use a rototiller or spade to cut each plant row to a 6-inch width. This sounds drastic, but do it! Runner plants from the 6-inch strip of "mother" plants will form a new matted row of plants. Apply 5 pounds of 5-10-5 fertilizer per 100 feet of row (10 pounds if sawdust is used as a mulch). Apply about 1-1/2 to 2 inches of water a week to the newly forming bed during June, July and August (include rainfall in total). To prevent overcrowding of plants and reduce the incidence of leaf diseases, thin plants to 5 to 7 plants per square foot. This method of renovation can be used to fruit the strawberry bed for a total of three or four years. During the rest of the growing season, irrigate to provide 1 inch of water per week, and continue to control weeds.

- Replant beds after harvesting berries for three or four years. After several years, strawberry plants tend to become infected with several leaf diseases and have declining yields. When this occurs, plants infected with diseases may be removed and a different crop can be planted on this site. Establish a new strawberry bed next spring where another crop was previously planted. To avoid infecting new strawberry plants with verticillium wilt, do not plant them where tomatoes, potatoes, peppers, eggplant, okra, raspberries or vine crops were planted within the past three years.

Diseases

Black Root Rot

Caused by soil fungi, nematodes, winter injury, fertilizer burn, drought, too much water or any combination of these. Plants show poor vigour, produce few runner plants and may die in patches. Roots may be spotted with black zones or lack feeder roots, or the entire root system may be killed. Infected roots are black when cut in a cross section. There is no way to control black root rot. However, proper cultural practices that minimize plant stress should limit this disease. If a bed becomes infected with black root rot, plant a new bed in different location, and use only healthy, white-rooted plants.

Red Stele Root Rot

Caused by a soilborne fungus and occurs most frequently on poorly drained sites with clay soils. The fungus is most active in the spring and fall when soils are cold and wet. Small feeder roots are killed, giving a "rat tail" appearance. Main roots that are infected have rusty-red colour in the core or stele. Plants may wilt and die during hot, dry conditions. To avoid this disease, plant only red-stele-resistant varieties and plant in a well-drained site in a loamy soil.

Leaf Spot and Leaf Scorch

Two common foliage diseases. Leaf spot produces spots 1/8 to 1/4 inch in diameter with purple borders and light gray centres. Leaf scorch produces small solid purple spots with rather undefined borders. When numerous they almost cover the entire leaf. Cultivars vary considerably in tolerance to these diseases. Foliage fungicides aid in the control of these diseases.

Verticillium Wilt

Caused by a soilborne fungus. In new strawberry plantings, symptoms appear about the time runners start to grow. In established plantings, symptoms appear when the berries ripen.

Older outer leaves of infected plants wilt, turn brown around the edges and between the veins and eventually die. Petioles of the leaves may also turn red. Few new leaves develop and new roots that grow from the crown are short and have blackened tips. To avoid this disease, select resistant strawberry cultivars and avoid sites that were previously planted with wilt-susceptible crops.

Gray Mold

Caused by the fungus *Botrytis,* which overwinters on plant debris on the ground. During bloom the fungus may be spread to blossoms during rainy weather, and the infection then spreads to the developing fruit. This disease often occurs where a dead petal or leaf adheres to the fruit or where a berry touches the soil. The disease appears as a gray, fuzzy mass of fungal spores on berries. To prevent this disease, thin plants to improve air circulation within the bed and use straw mulch. Fungicides may also be applied.

Leather Rot

Caused by a fungus that infects the berries when ripening fruit clusters are in contact with the soil or standing in water after a period of warm, rainy weather. Infected immature fruit turns brown. On ripening fruit, the colour becomes bleached out and infected areas are tough and dry. Berries taste bitter and smell rotten. A single infected berry can ruin the flavour of a whole jar of jam. To prevent this disease, use a mulch to minimize rain splash onto the fruit, or fungicides may be applied.

Viruses

May stunt plant growth or cause mottling or curling of leaves. They are introduced by insects such as aphids or leafhoppers. To prevent viruses, use certified plants and control these insects.

Insects and other Pests

Tarnished Plant Bugs

Cause deformed or "nubbin" berries with a concentration of seeds at the tip of the fruit. Nymphs puncture individual seeds and inject a toxin so that the fleshy part of the berry stops developing where the seed was injured. Damaged seeds are hollow and turn a straw-brown colour. Plant bugs are often in alfalfa; to prevent the bugs from feeding on strawberries, avoid mowing alfalfa when strawberry plants are in bloom.

Leafrollers

Small worms that roll the leaves together and feed on the leaves. They must be controlled if they become numerous.

Mites

Tiny, spiderlike creatures that are typically found on the underside of leaves. Mites tend to be the most damaging during hot, dry periods, when they suck plant juices from the leaves.

Strawberry Weevils or Clippers

Puncture fruit buds with their snouts to feed on immature pollen. Later, females deposit an egg inside a floral bud and girdle the petiole of the bud so that it falls to the ground or is left hanging by a small bit of tissue.

Slugs

Eat holes in ripe fruit and are worse during rainy weather or near rotting foliage. Shiny slime trails are evidence of slugs.

Nematodes

Found in the soil and are too small to be seen with the naked eye. Most nematodes feed on roots. Symptoms of nematode damage include galls on the roots or stubby, stunted roots. To avoid a buildup of nematodes, rotate plantings to a different site when establishing new beds.

Causes of Malformed Fruit

"Nubbins" or "button berries" may be caused by feeding tarnished plant bugs, heavy infestations of cyclamen mites, frost injury, nutrient deficiencies, inadequate pollination, abnormally high temperatures that make pollen nonviable, or by the application of 2,4-D from Aug. 1 to Nov. 1, when fruit buds are differentiating.

Fasciated berries, those that have a cockscomb shape as if several berries have grown together with multiple tips, are caused by short day-lengths in the fall or by cold, dry weather during this time. Fasciated fruit often occurs when cultivars adapted to northern conditions are grown in the south.

Tree Fruits: Organic Production Overview

This publication focuses on production and marketing of organic fruits and nuts, highlighting a systems approach to orchard production and farm management. Not intended as a comprehensive production guide for individual fruit crops, this publication introduces key production issues that merit consideration in any specific crop or production region. General information on cultural practices for fruit production (choosing varieties, spacing, pruning, training, irrigating, harvesting, postharvest handling, etc.) is relevant to both organic and conventionally managed operations, and it is widely available from the Cooperative Extension Service, nurseries, and in horticulture literature. A note about Organic Standards: Included in this publication are references to the organic standards authorized by the USDA's National Organic Program, www.ams.usda.gov/nop. Organic producers should verify with the appropriate certification bodies that their practices and any materials they intend to use are compliant with applicable standards for their intended markets. This is especially true if those markets are international, where there may be additional production and labelling requirements.

Organic Fruit Production

Organic production is defined by USDA's National Organic Program (NOP) as "A production system that is managed...to respond to site-specific conditions by integrating cultural, biological, and mechanical practices that foster cycling of resources, promote ecological balance, and conserve biodiversity."

Organic fruit production involves more than simply excluding synthetic pesticides and fertilizers. Benign neglect does not meet NOP standards for production. Organic agriculture is an integrated approach to active and observant management of a farming system. It begins with good soil management for nutrient cycling, productivity, and tilth. It involves an integrated, preventative approach to pest management to protect the health and productivity of the orchard.

Marketing and Economic Considerations

To plan for economically successful enterprises, farmers must design their fruit production systems to match their marketing strategies. Good fruit production alone does not lead to a successful enterprise. Profitability depends on a combination of production volume, quality, size, and a reliable marketing strategy. Marketing channels range from direct markets to wholesale shippers. Growers must understand what each of their customers wants and be prepared to meet the expectations of the markets they intend to reach. For example, at farmers' markets, customers seek good tasting fruit at or near the peak of ripeness for prompt consumption, but supermarket distributors demand that fruit be uniform and shippable. It is important to market in an appropriate niche, one where the production of your operation can consistently meet the buyers' expectations of volume, quality, and timing.

Good production and good quality do not guarantee good returns.

My late friend, farmer, and master machinist Grover Parsons recounted how he planted 10 acres of apricots in San Juan Bautista, California. He chose the best-tasting variety, Royal Blenheims. He cultivated them organically, but that was before there were organic standards for market differentiation and premium pricing.

The variety was also well-suited to the climate and soil. They grew beautifully and began producing good yields of excellent fruit. He harvested the apricots at the appropriate time and sold them to a local buyer. With each successive season he found the price more and more discouraging; it did not adequately compensate him for his work. He got on his tractor and took out the orchard—all but a few trees along the fenceline. Those trees were still providing him with good fruit for home use decades later.

Ann Baier

Premium pricing can be critical to the viability of organic fruit operations, because production costs are often higher than those for conventional orchards. Organic pest control, particularly labour costs for hand thinning and weed control, is generally more expensive than conventional practices. Yield and quality can vary widely, depending on the growing season and management practices. In the past it may have been true that organic yields and pack-out rates (the percent of marketable fruit) were lower than in conventional production. Today, however, those differences have narrowed, and yields in some organic production systems can match or exceed those of conventional systems. To achieve good yields, organic growers must be prepared to develop innovative production and marketing strategies. Many commercial organic fruit producers, especially family-scale farmers, minimize waste and losses of potential revenue by processing (drying, preserving, or juicing) fruit considered unsuitable for the fresh market.

There are tradeoffs in every marketing strategy. A successful grower must develop markets in which the price for organic produce adequately compensates for all production costs. Additionally, the marketing process must be compatible with the grower's personality and business skills. The particular combination of components in any grower's marketing strategy will depend on local marketing opportunities as well as the grower's desire to be directly involved in marketing, tolerance for stress, and ability to balance a variety of risk factors.

Cost-Benefit Analysis of Markets

Several certified organic apple growers in the Central Coast region of California sell their culls to Martinelli's, a longstanding locally-based juice company. The company has a low-volume organic juice line that produces a limited number of runs each year, if farmers can time their harvest to match it. Even though the company does not give a premium for organic fruit that goes into their regular juice line, its proximity to the farms, fair payment practices, and good business relations make for a satisfying and cost-effective sale for many organic growers. Although there are organic processors in other parts of the state, the price premium for organic fruit does not always offset the increased transportation costs, volume requirements, hard bargaining, or worry about late payments from more distant organic processors. Being certified organic increases one's marketing flexibility. An organic certificate provides the option but not the obligation to use the organic claim. When it is advantageous, organic produce can be sold as conventional, but never vice-versa.

Cultural practices in fruit production begin with selection of an appropriate site, fruit crop, rootstock, and fruit variety, followed by site preparation (tillage and pre-plant soil amendments) and orchard layout (tree and row spacing). These considerations will largely determine the productivity, health, and efficiency of operations in your orchard over the long term, and they cannot easily be changed once the orchard is established. If you are managing an existing orchard, you will continually need to take stock of its assets and limitations in relation to current markets, and either work within those limits or make plans for some degree of orchard renewal. If you are considering the purchase of an existing orchard, ask the owners about their financial history (production costs and sales records), and research the market to assess the economic feasibility of continuing with the business as it is currently practiced. Once you have made the most realistic cost estimates possible, you can develop a plan to adjust the production system, revise the marketing plan, or walk away while you still have your shirt.

Planning and Planting an Organic Orchard

Site Selection

Fruit trees, like most crops, respond to good soil with vigour and productivity. Trees can successfully produce economic yields on hillsides, rocky soils, and other sites not suitable for frequent tillage. Look carefully at your site and take stock of its soil, slope, and aspect, water infiltration and drainage, frost patterns, maximum and minimum temperatures, length of growing season, distribution of annual precipitation, availability of water for irrigation, proximity of the water table, and wind and air circulation patterns. Most of these are beyond your control, and your planting plan must suit the natural conditions of the site. While farmers may be able to improve the soil over time, they cannot change the subsoil layers, influence the prevailing wind, or modify temperatures to any significant extent.

All the factors regarding site suitability for conventional fruit plantings apply—even more so—to organic operations. While conventional growers may fall back on chemical fertilizers and pesticides to compensate for some poor site decisions, organic growers cannot. Good drainage and air circulation are essential for disease control. The presence of certain weeds and forage species is of particular concern to the organic grower. Bermuda grass, Johnson grass, quack grass, and several other pernicious species can be serious problems to fruit growers and are difficult to control with organic methods once an orchard is established.

An assessment of physical and environmental factors will help the grower determine whether a crop can be grown easily, marginally, or not at all. While someone with a home orchard may consider it worthwhile to cover a lemon tree before each anticipated frost, or to nurse a few apricot trees through Midwestern winters in order to savor the delicious fruits two seasons out of seven, these would likely not be viable commercial enterprises. However, stretching the limits of production within reason can be worthwhile under certain circumstances. For example, subtropical fruits grown in the coastal valleys of California bring a premium for their freshness and novelty, offsetting the expense of the extra care they require. Depending on the crop, harvesting either early or late in the season can also provide a market advantage. While California's San Joaquin Valley is not known for apple production, its warmer spring and summer temperatures can bring the crop to maturity a few weeks ahead of coastal producers. The price premium for first-of-the-season organic fresh-market apples may offset the overall lower crop yields. Fruit grown in its primary growing region may be more difficult to distinguish from the rest of the fruit in the market, and so lose its competitive edge.

Pragmatism is critical in crop and variety selection. The fruit grower must take into account not only factors of yield, productivity, quality, and flavour, but also of marketability. Diversification of varieties and marketing channels is a prudent strategy. While one farm in California may receive a good return for the fruit from one lone jujube tree, there is no assurance that such a profit can be scaled up to an extensive orchard of speciality fruits. Conclusions: while the jujube is a tasty and unusual fruit, and the value of fruit from one tree may be substantive, the regional market for this novelty would not accommodate production from 200 trees. Similarly, in his book *Epitaph for a Peach,* Mas Masumoto describes the wonderful flavour and quality of the Suncrest peach. Yet his family farm substantially decreased their plantings of this variety. It was a practical business decision to plant more of the newer varieties that the current market demands, and keep a smaller block of the sumptuous Suncrests that are more difficult to ship and store. This seasoned fruit grower advises a production and marketing plan that includes a combination of mainstream and speciality varieties.

Fruit Crop and Variety Selection

Because fruit trees are perennial and represent a considerable investment of both time and money, it is important to start by planting your orchard with the optimum varieties for your location and intended markets. Research on the front end can pay the grower back many times over. Information on species and varieties is available from Cooperative Extension, nurseries, and other local growers. Many land grant universities have field stations where they have planted many varieties of fruit trees and gathered data and observations over several years. A visit to such a site can provide you with the invaluable opportunity to see the trees growing, talk with the manager of the experiment station about production challenges such as pests and diseases, and even taste the fruit.

Crop Species Selection

Clearly, the first decision is what species to plant. Is a tree orchard the best use of your land and talents? Or is your site and marketing plan better suited for a somewhat shorter-term investment in smaller plants such as blueberries, caneberries (raspberries, boysenberries, olallieberries, other blackberry varieties), grapes, kiwi, or even strawberries? If you are sure that you are willing to manage tree fruits and nuts, will your focus be to produce almonds, apples, apricots, avocados, cherries, figs, grapefruit, jujubes, lemons, oranges, pawpaws, peaches, pecans, pears, persimmons, plums, pluots, or zapotes?

Careful consideration of environmental conditions, as well as the locations of markets and suppliers, is of tantamount importance. For example, organic peach production in the East is greatly complicated by the presence of the plum curculio and by greater disease pressure than in the drier climates of the West. In general, the West's arid climate is better for organic fruit production. The small fruits (blueberries, blackberries, raspberries) are easier to produce organically than tree fruits in almost all locations.

The availability of production supplies and markets in your region can be a critical factor in crop selection. Being the only one growing a certain fruit may provide you with a local marketing niche; however, the value of readily available supplies and services should not be underestimated. While some supplies can be easily and cost-effectively shipped by mail, others cannot. Pest management materials such as codling moth pheromone traps can be efficiently shipped from a distant supply company. But how far do you have to drive to purchase boxes and bulky packaging supplies? How far to cold storage, a packing house, distributor, processor, or transportation terminal? Driving several hours to purchase appropriate boxes or to deliver fruit to a broker's cooler can make an otherwise viable enterprise unprofitable.

Variety and Rootstock Selection

Once the question of crop species is settled, the next decision is what variety (or combination of varieties) to plant. Considerations include, but are not limited to:

- harvest season: early, mid, or late season, or a combination of these to achieve a more continuous supply or to ensure a crop during early or late marketing windows
- adaptability to the region: cold hardiness, temperature ranges for optimal growth, requirements for soil fertility or pH
- chill requirements for fruit set and flavour

- water requirements: need for irrigation or protection from waterlogging
- stature: dwarf, semi-dwarf, or standard
- resistance to diseases and pests
- marketability: colour, flavour, nutritional value, storage requirements, shipability, uniformity, shelf life—any characteristics that define quality for your customer
- proximity to appropriate markets

You can select for desired characteristics, especially in grafted trees, with a combination of varieties of rootstock and fruiting wood.

Sources of Planting Stock

It is important to get clean planting stock. Buying from reputable nurseries that provide stock certified by state inspectors to be free of diseases and insect pests is best. Organic planting stock is required, if commercially available, for certified organic fruit production. If organic planting stock is not available, organic growers must document their search for organic stock and its lack of commercial availability. Most certifiers interpret the organic standards as requiring organic management of non-organic planting stock for at least 12 months before harvesting a crop that is to be sold as certified organic. With newly planted tree crops, this is a non-issue, since they generally grow for at least three years before producing a marketable crop.

Type and Size of Planting Stock

The type of rootstock—standard, dwarf, or semi-dwarf—will determine the size of the tree at maturity. Tree size determines the spacing, number of trees per acre, training system, years to bearing, and timing of economic return. Orchard design should reflect the grower's production and cash-flow goals. For example, standard trees produce more fruit when mature, and initial purchase and planting costs are lower. Smaller trees have higher initial planting costs, since more trees are needed to achieve density.

Dwarf and semi-dwarf trees generally come into production sooner. Smaller trees simplify many field operations, including pruning, grafting, thinning, pest management, and harvest. Efficiency and safety are greater when a majority of operations can be accomplished from the ground as opposed to on ladders or by climbing. Weeds are less of a problem in the shade of a densely planted orchard. Depending on the species and variety, bareroot trees are often the most practical form of planting stock to ship, and the most economical to purchase. This is a good option for deciduous trees. Other varieties, such as citrus, must be purchased in containers. Given the option of different sizes of bareroot trees, some walnut growers say that investing in a 1-inch tree over a 3/4-inch tree is worthwhile, because larger trees grow more vigorously. An experienced apple grower who produces without irrigation beyond the first year, however, stated his preference for 5/8-inch bareroot trees, which have a good balance of roots and are neither too big nor too small.

Disease and Pest Resistance

Genetic resistance refers to inheritable traits that enable a plant to inhibit disease and resist pest damage. A very important control measure for organic growers is to choose cultivars that are resistant to the pests – especially the diseases – most prevalent in their areas. In some cases, such as that of bacterial spot in peaches, cultivar resistance is the best or only control for a particular disease. A cultivar may be quite resistant to one disease but still susceptible to another. Prima apples for instance, are very resistant to scab but very susceptible to cedar-apple rust. A planting stock resistant to a particular pest provides only relative resistance, not absolute immunity. A moderately resistant or tolerant variety may show symptoms of the disease but exhibit little to no reduction in yield.

Disease resistance must be weighed against other advantages. For example, walnut growers in the coastal regions of California have lost large numbers of trees in recent years to "black line," a fungal disease for which there is no treatment, only resistance. Payne variety is susceptible and Chandler is highly resistant to this disease. (Some say that Chandler also appears to suffer lower damage levels from codling moth and walnut husk fly, the other common walnut pests in the region.) A trade-off is that Paynes mature sooner and can be harvested earlier in the fall, whereas Chandlers come in at least a month later when early rains can hinder harvest operations and make field preparations for planting a winter cover crop difficult or impossible.

While no fruit trees are resistant to insects that damage their fruit, it is possible to find stock that is resistant to insects that feed on other parts of the plant – *Phylloxera*-resistant grape rootstocks, woolly aphid-resistant apple rootstocks, and nematode-resistant peach rootstocks, for example.

As important as this resistance is, there is no cultivar of any fruit species with multiple insect pest resistance; therefore, an integrated pest management plan is necessary to protect fruit plants from a complex of several pest species. It will be important to identify the most troublesome pests for your crop and region in terms of frequency of incidence, severity of damage, cost of control, and economic consequences of the damage. Then seek out varieties that are resistant to those key pests and take into account any trade-offs you may make with other desirable characteristics, including seasonality, productivity, and flavour. Substantial crop-and variety-specific information on pest and disease resistance is available on the Web site of the University of California IPM project and other university pomology departments. Be sure to check with local suppliers of planting stock, and talk with other growers in your area about what has worked best for them.

Site Preparation

Important considerations in site preparation include alleviating soil compaction, enhancing fertility, adjusting soil pH, and managing weeds, pests, and diseases. Attention to the details of site preparation can help reduce weed and disease problems and assure a vital planting through soil improvement. What needs to be done depends on the previous use of the land, including crops grown, current vegetation, and the presence of pests and diseases. Many growers rip or chisel the soil to loosen layers of compaction before they plant a new orchard, since deep tillage will be disruptive once the trees are established.

Before establishing an orchard, it is important to adjust the soil pH to best suit the crop you've selected. Soil tests can assess current soil conditions, including pH, mineral levels, and their relative proportions. Traditionally, pH has been adjusted through applications of lime (to raise the pH) or sulfur (to lower pH). Most fruit plants perform best around pH 6.5, although they tolerate a pH range between 5.5 and 7.2. Blueberries are an exception. They require an acid soil—ideally pH 4.8 to 5.2. Soil test results help to guide applications of soil amendments such as compost, lime, gypsum, or other rock powders, to provide good soil conditions that meet the nutritional needs of the orchard.

In general, fruit crops do not require highly fertile soils for good production, though this varies with the species. Highly fertile soils, rich in nitrogen, can promote too much vegetative growth at the expense of fruiting in trees such as apples. A nutritionally balanced soil, proper soil pH, and plentiful organic matter are the fundamentals of an organic fertility management plan for fruits. Pre-plant soil improvement for organic fruit plantings usually involves some combination of cover cropping and applications of compost, natural minerals, or other organic fertilizers.

Weed Management Prior to Orchard Establishment

It's easier to manage weeds before an orchard is established. Cover crops produce a thick stand that will shade or choke out weeds. Combined with a well-planned sequence of tillage, cover cropping is an effective pre-plant weed suppression strategy that also contributes to soil fertility and stable humus. The basic strategy begins with plowing under or disking the existing vegetation, ripping or deep chiseling to loosen compaction, planting a cover crop to suppress weed growth, mowing down and tilling under the cover crop(s), and finally planting the fruit crop. Several cover crop and tillage sequences may be necessary before planting.

Specific cover crops and management strategies vary with location and purpose. The two cases below raise the kinds of questions you need to ask to choose an appropriate cover cropping system. The cover crops you choose for site preparation (before planting the orchard) may be entirely different from those you want once the orchard is established.

Bart Hall-Beyer, co-author of *Ecological Fruit Production in the North* (1), provides one example of how cover crops can be used to suppress weeds in the growing season prior to fruit crop establishment. His program consists of fall plowing, to allow the sod to rot, then disking as soon as the soil is dry in the spring, followed by harrowing every 10 days for at least one month to kill germinating weeds. He next incorporates compost and mineral nutrients and seeds buckwheat as a smother crop. He then tills the buckwheat into the soil after it has started flowering but before seed-set. Hall recommends additional cultivations at 10-day intervals, followed with rye as a fall cover crop. The rye is incorporated the following spring and the fruit crop planted.

In the Mid-South, researchers at the Kerr Centre for Sustainable Agriculture in Poteau, Oklahoma, evaluated a number of cover crops for weed suppression on heavy soils. They converted pasture land to horticultural production, using rotations of cover crops and tillage. By this method, they virtually eradicated Bermuda grass from the fields in one to two years. Among their general observations are the following.

- Dense warm-season cover crop plantings of several species demonstrated a high degree of weed suppression, whether close-drilled in 6-inch rows or planted on wider 32-inch rows and row-cultivated.
- The length of the warm season may allow more than one cover crop to be grown in succession. Some cover crops may also be cut and allowed to regrow.
- Legume cover crops of purple hull peas (cowpeas), crotolaria, and sesbania all demonstrated good-to-excellent weed suppression, while supplying nitrogen and biomass to the soil.
- Of these, sesbania produced the most biomass and was the most effective weed suppressant. When cut with a sickle-bar mower at flowering, it regrew well and continued to suppress weeds. It is very drought-tolerant. Seed cost and delivery, however, were quite high. If allowed to re-seed, sesbania can create a moderate weed problem the following year.
- Crotolaria (sun hemp) was a better nitrogen producer, but a less effective weed suppressant than sesbania. It, too, can be cut at flowering with a sickle-bar mower and allowed to regrow. Like sesbania, it is very drought-tolerant. The cost of seed can be high. Crotolaria seed is toxic—especially to birds—and the plants should not be allowed to go to seed.
- Because crotolaria and sesbania are quite fibrous, they should be mowed with rotary or flail mowers prior to soil incorporation.
- Cowpeas produce somewhat less nitrogen than crotolaria, and less biomass than either sesbania or crotolaria. They are, however, less fibrous and, therefore, decompose faster.
- Allowing cowpeas to flower and produce mature, dry seed prior to incorporation creates an inexpensive, self-seeded succession cover crop.

- Sudan grass proved the most effective of all warm-season weed suppressants. It can be flail or rotary mowed several times if regrowth is desired.
- Winter cover crops can be planted in rotation with warm-season cover crops. A combination of grain rye and hairy vetch was the most effective in this location. Winter peas and oats, and winter wheat—often grown in combination—also have good competitive ability.

Soil Solarization

Soil solarization is placing transparent plastic films on moist soil to capture solar energy. Solarization takes four to eight weeks to heat the soil to a temperature and depth that will kill harmful fungi, bacteria, nematodes, weeds, and certain insects in the soil. Solarization can be a useful soil disinfestation method in regions with full sun and high temperatures, but it is not effective where lower temperatures, clouds, or fog limit soil heating.

Other disadvantages of solarization as a weed control method include its expense and disposing of the plastics. Solarization is most commonly used in smaller areas, such as greenhouses and nursery beds, though it has been used experimentally to treat orchard soils, either prior to planting or during establishment. Experiments are underway to evaluate using biodegradable spray mulches for solarization.

Researchers emphasize that solarization should be seen as just one component of an integrated pest management system, rather than as a "stand alone" technology. A University of California Cooperative Extension leaflet (No. 21377), "Soil Solarization: A Nonchemical Method for Controlling Diseases and Pests," details the technique.

Orchard Layout and Design

Orchard layout influences the long-term health of the trees and the ease of field operations such as pruning, irrigation, fertilization, and weed and pest management. Everything is related: the decisions you make about the space between rows and between trees in the row will have an impact on everything from disease management to harvest operations. While the specific spacing and training of trees will largely depend on the species, the following questions offer general considerations that will save time, resources, and expenses throughout the life of the orchard.

What is the lay of the land? Which way does the water run? What is the angle of the sun during different seasons? How will these affect the movement of both water and air, and in turn, temperature and humidity levels, crop ripening, and incidence of diseases and pests? Do the rows need to be planted on the contour for soil conservation or to capture limited seasonal moisture? Or should they be sloped to drain excess moisture? Given the degree of slope, which direction will provide the greatest safety for operating equipment and ease of harvesting ?

What are the diseases and pests that affect this crop in this region? What are their life cycles? Alternate hosts? Natural enemies? What conditions favour their growth and severity? What design strategies might promote or reduce these conditions? Would a certain orientation of the rows provide better exposure to the sun or better air circulation? Will you rely on seasonal pruning to maintain an open canopy to increase air flow through the foliage and sun to the fruit?

What equipment will you use for field operations? Consider all possible tasks, including planting, mowing (or incorporating orchard floor vegetation), cultivation, pruning, irrigation, application of materials for pest management, and fruit harvest. Be sure that your row spacing is adequate to allow entry and maneuverability of any tractor, trailer, spray equipment, string trimmer, wagon, wheelbarrow, or hand cart that you plan to use. What crop density do you seek? How soon after planting? The decision will depend on the species and stature of your trees, the cost of purchasing and planting them, the years to maturity, the prevalence of weeds, and other considerations. Using close in-row spacing or double rows of trees may complicate weeding in the first year or two, but thereafter shading will greatly reduce the need for weeding the inter-row. Some farmers plant slower-growing trees (such as walnuts) using closer spacing, then remove every other tree when they reach a certain maturity. The estimated benefits of earlier harvests must be considered against the costs of planting, managing, and eventually removing the trees. Alternatively, annual crops can be grown between immature orchard trees.

Managing an Established Organic Fruit Orchard

Orchard Floor Management/Cover Crops

The orchard floor—the tree rows and alleyways— can be managed in a variety of ways, using tillage or mowing with cover crops, grazing, or mulching (described in detail in the Organic Weed Management section). A system that provides full ground-cover provides the best protection against erosion. Some fruit growers have practiced "clean cultivation," eliminating vegetation throughout the orchard, but this system has many disadvantages, even if accomplished with allowed tillage practices instead of organically prohibited herbicides. A bare orchard floor is prone to erosion, gradual depletion of organic matter, increased soil compaction, and reduced water infiltration. It's also difficult to move equipment through the orchard in wet weather. However, a ground cover that is actively growing in the summer uses up water. This is a severe disadvantage in irrigated orchards where water is limited and expensive.

Orchard floor management can control erosion, improve the soil, and provide beneficial insect habitat.

- Where they are adapted, orchard grass, fescue, and other cool-season grasses are practical because they go dormant during the heat of the summer, minimizing competition with the fruit crop for water. With proper fertility management, these grasses can also provide plentiful mulch. Likewise, grasses are a good choice in apple orchards, for example, where the excess nitrogen provided by legumes can actually reduce fruit yields.

- Many warm-season legumes are deep-rooted and compete with the trees for water. Normally, they should not be allowed to grow under the tree canopy. However, leguminous ground covers can provide significant nitrogen to fruit trees or vines. Grass and legume ground covers alike promote water infiltration and hold the soil in place during the rainy season. Ground covers help maintain and increase soil organic matter, which increases the soil's ability to retain moisture. Cool season legumes, such as fava or bell beans, vetches, and clovers, also can achieve these goals.
- Planting subterranean clover into established orchards can provide mulch, fertilizer, between-row ground cover, and beneficial insect habitat. This clover reseeds itself in early summer and dies back during the hottest part of the growing season, leaving a relatively thick, weed-suppressive mulch. This system is used in apple and peach orchards in Arkansas and for a variety of orchard crops in California, but not where winter temperatures regularly drop below 0° F. Subterranean clover can provide habitat for such beneficial insects as ladybeetles, syrphid flies, big-eyed bugs, soft-bodied flower beetles, and other predators.

Crop Rotation

In an organic orchard, crop rotation does not mean changing the economic crop itself, but diversifying the vegetation that grows around the fruit crop. California organic almond farmer Glenn Anderson describes how important maximum plant diversity is within the orchard and in the surrounding vegetation.

He takes advantage of every practical opportunity to diversify vegetation: the orchard floor grows cover crops; the landscaping around the family home situated in the midst of the orchard provides shelter and food for a variety of beneficial species; the roadway, farm perimeter, and even the paths of the irrigation lines provide habitat for these beneficials.

Research studies confirm the positive effects of organic practices on beneficial insects. Several articles compared yields, pest and beneficial insect populations, and water and air quality factors on Anderson's farm to those of his brother's adjacent, conventional almond farm and found favourable results with organic practices. Mr. Anderson gives credit for the health of his orchard to the host of creatures that contribute to ecological balance on the farm. He believes that all the trees, shrubs, and plants he encourages help sustain beneficial insects, spiders, bats, and birds within and around the orchard. The USDA National Organic Program Final Rule defines crop rotation in 7 CFR § 205.2 as "The practice of alternating the annual crops grown on a specific field in a planned pattern or sequence in successive crop years so that crops of the same species or family are not grown repeatedly without interruption on the same field.

Perennial cropping systems employ means such as alley cropping, intercropping, and hedgerows to introduce biological diversity in lieu of crop rotation." According to 7 CFR § 205.205 Crop Rotation Practice Standard, the benefits of rotation are to :

"(a) Maintain or improve soil organic matter content;

(b) Provide for pest management in annual and perennial crops;

(c) Manage deficient or excess plant nutrients; and

(d) Provide erosion control."

Cover Crops

Steps and considerations for selecting and managing a ground cover:

- State your objectives in order of priority. For example: suppress weeds, break up soil compaction, add organic matter to the soil (increase tilth, water infiltration rates, and water-holding capacity), enhance soil fertility (fix nitrogen), attract and sustain beneficial insects, serve as a trap crop for pests.
- Take into account the climate, rainfall pattern, soil type, and potential for soil erosion.
- Describe desired growth patterns and characteristics: Does this cover crop have a tap root? Will it regrow if mowed? Does it fix nitrogen? How much biomass does it produce? Is it fibrous? How long will it take to break down? Will I need to mow or chop it to speed its decomposition? When should I incorporate it? Will it reseed itself? What is its potential to become weedy if it goes to seed? Does it attract insects? What kinds? Will it serve as beneficial insect habitat? Is it a host for pests? Can it be used as a trap crop?
- Consider planting techniques and timing: When and how should I plant a cover crop? How can I manage its growth for production of organic matter and nitrogen fixation? Are there seasonal weather constraints to getting equipment into the field? What methods provide the best germination rate for the effort—broadcast, drilled, frost-seeded (the technique of broadcasting seed so that it is incorporated by the motion of the soil freezing and thawing)? What equipment do I have available—disc, broadcast seeder, seed drill, flail mower, chisel plow, spading machine? What is the seed cost? Do I need to inoculate it with *Rhizobium* bacteria to increase nitrogen-fixing nodulation?
- Is the best cover crop for my situation a single crop, a mixed seeding, or a series of different cover crops?

Pest Management

Organic pest management relies on preventative cultural, biological, and physical practices. Organisms—insects, mites, microorganisms, or weeds— become pests when their populations grow large enough to prevent growers from reaching production goals. Integrated Pest Management recognizes that the mere presence of a potentially damaging species does not automatically mean that control actions are necessary. Knowledge of pest life cycles and monitoring techniques developed in IPM programs are useful for organic growers as well, because they mirror some of the elements of the organic pest management standard.

The National Organic Program Final Rule 7 CFR §205.206 Crop Pest, Weed and Disease Management Practice Standard reads:

(a) "The producer must use management practices to prevent crop pests, weeds and diseases including but not limited to:

1. Crop rotation and soil and crop nutrient practices...;
2. Sanitation measures to remove disease vectors, weed seeds, and habitat for pest organisms; and
3. Cultural practices that enhance crop health, including selection of plant species and varieties with regard to suitability to site-specific conditions and resistance to prevalent pests, weeds and diseases.

(b) Pest problems may be controlled through mechanical or physical means including but not limited to:

1. Augmentation or introduction of predators or parasites of the pest species;
2. Development of habitat for natural enemies;
3. Non-synthetic controls such as lures, traps and repellants.

(c) Disease problems may be controlled through

1. Management practices which suppress the spread of disease organisms; or
2. Application of non-synthetic biological, botanical or mineral inputs.

(d) When the practices provided for in paragraphs (a) through (d) of this section are insufficient to prevent or control crop pests, weeds and diseases, a biological or botanical substance, or substance included on the National List of synthetic substances allowed for use in organic crop production may be applied to prevent, suppress or control pests, weeds, or diseases: Provided, That the conditions for using the substance are documented in the Organic System Plan."

Three tiers of pest management strategies are described in the NOP Final Rule.

1. First, the producer should use cultural management practices that prevent pest and disease problems. These include multiple components of a holistic, systems approach to organic farm management and crop production.

2. In the second tier of pest management, biological and physical methods provide additional protection and need no justification. These practices build on and complement good cultural practices, but cannot compensate for poor cultural practices.
3. The third and final tier—the last resort—may be applying an allowed material if the first two tiers of response are ineffective and if the conditions for their use are described in the grower's Organic System Plan (OSP). A material response may be necessary under some circumstances, but it will be just one component of an integrated pest management plan that is part of an overall OSP.

Mexican Fruitfly on Grapefruit

Common arthropod pests of fruits include insects (aphids, caterpillars, leafrollers, twig borers, flies, psylla, scale insects, leafhoppers, mealybugs, earwigs, thrips, and beetles) and mites. Identification and preventative management are essential to organic production systems. Identification charts are available from many university Extension Web sites and publications. Organic Fruit Tree Management provides a useful list of important fruit pests, their hosts, status (potential impact on various stages of fruit tree growth as well as on crop production and quality), identification (adult, immature/ larva, pupa, and eggs), life cycle, monitoring/ thresholds, and management. This book was written for fruit growers in the North and may not address the key pests in all other regions. Nonetheless, its approach and outline serve as an extremely useful models for growers developing informed and integrated organic pest management strategies for their orchards. While there are many other components to insect and mite pest management, in recent years there has been a good deal of research on vegetation management to enhance natural biological control. Approaches to cover crop and vegetation management described by Bugg and Waddington include:

- resident vegetation that harbours beneficial arthropods (insects, mites, spiders);
- strip management of cover crops to ensure the continuous presence of habitat for both beneficials and pests;
- insectary mixes of plants attractive to beneficial arthropods;and
- use of mulch from mowing to harbour generalist predators.

There is also increasing evidence that managing vegetation adjacent to economic crops (fencelines, roadsides, etc.) as habitat for beneficial insects has a positive impact on pest management. These areas often include native plants and shrubs that flower at different times of the year, providing sources of pollen and nectar for beneficial arthopods.

The long-term nature of growing fruit using cover crops and other resident vegetation management can sustain populations of predators, parasites, and other beneficial organisms. There are many possible trade-offs that emphasize the need for careful planning and the importance of research and monitoring.

- Apples and pecans: California apples and Georgia pecan orchards planted in a diverse mix of cover crop species provided habitat and food for an array of beneficial organisms, resulting in a decrease of orchard pests.
- Peaches: Some winter annual broadleaf weeds have been implicated in increased populations of tarnished plant bugs in peach orchards, and dandelions and chickweed can serve as hosts for viruses that affect peaches and apples.
- Walnuts: Two species of ladybeetles were more abundant in an orchard floor where a cover crop was maintained from February to May, and helped keep walnut aphid populations in check.
- Apples: Codling moth infestations of apples were lower where bell beans grew. (Bell beans are known for their extrafloral nectaries that help sustain beneficial insect populations even when the flowers are not open. Insect-eating birds can also reduce codling moth populations, but not control them. The development of pheromone mating disruption has been a major breakthrough in the past several years, making organic codling moth management feasible and organic apple production competitive.
- Cherries: In regions of California where the mountain leafhopper transmits buckskin disease, growers should use caution in establishing permanent covers that include cool-season alfalfa and clover species that harbour the leafhopper. This case emphasizes the importance of understanding and carefully considering the pest's life cycle, with respect to the presence of host plants where the pest can reproduce.
- Citrus and avocado: Wind-blown pollens from grasses and trees can be alternate food sources for the predatory mite Euseius tularensis in late winter and early spring, and may, therefore, help build and sustain populations of predatory mites that attack pest species that include the avocado brown mite, citrus thrips, citrus red mite, and scale insects.
- Some legumes are also known to attract hemipterous pests like tarnished plant bugs and stink bugs. Where these pests are a problem, legumes may be less desirable as orchard cover crops, unless they can be managed as trap crops for lygus bugs. Alternatives such as mustards, buckwheat, dwarf sorghum, and various members of the Umbelliferae (carrot, cilantro, dill, fennel, anise, etc.) and Compositae (sunflower and other composites) families support substantial numbers of beneficial insects without attracting as many pests. However, mustards flower and seed early, providing early season food for hemipterans, including stink bugs.

Conversations with several organic farmers reinforce these research findings. Many organic walnut growers plant cover crops that are mixtures of legumes—such as bell beans, vetch, or alfalfa—to produce nitrogen and create a beneficial insect habitat, in combination with cereals that produce organic matter and provide support for the legumes. Growers alternate rows when they mow or disc, intentionally leaving strips of cover crops in the orchard to provide areas with flowering plants that sustain populations of beneficial insects. In any orchard setting it is important to watch for gopher problems. In addition to their many benefits, cover crops can also provide food and cover for gophers.

Disease Management

Disease can be a significant limiting factor in organic fruit production. Diseases may be caused by fungi, bacteria, viruses, nematodes, mycoplasmas, or protozoans. Disorders caused by the weather or by nutrient imbalances (deficiency or toxicity) can create symptoms that look like diseases. Proper identification and preventative management are imperative. For example, boron toxicity or blossom-end rots cannot be cured with fungicides. Cooperative Extension and university Web sites can help in identification. The book *Organic Tree Fruit Management* includes detailed descriptions of diseases of fruit crops, including hosts, status (potential impact), symptoms, life cycle, monitoring/thresholds for treatment, and organic management strategies. This book's organic context and perspective are invaluable. Thresholds for action may be distinct in an organic pest management system (in which the impact of a disease organism is considered not only on this year's crop but also in terms of the future implications for control, potential buildup, and impact on the health of the trees), compared to conventional IPM programs (in which the availability and allowability of pesticides is assumed).

As described in the NOP standard, a combination of cultural controls forms the foundation for a good disease-management strategy. As discussed earlier in this publication, selecting resistant varieties or rootstock is of utmost importance, as is selecting the right growing location. In an established orchard, one can practice good sanitation by cleaning up debris, pruning, and removing diseased plants and disease vectors. Some plants can serve as alternate hosts for diseases. Eastern red cedars, for example, are alternate hosts for cedar-apple rust. Wild blackberries can harbour blackberry rust, and wild plums can foster peach brown rot. A good defence against plant disease is to maintain the crop plants in excellent health and vigour, with sufficient—but not excessive—soil nutrients and moisture.

Many diseases of fruit crops only affect a particular species and variety of fruit. There are, however, some diseases that are common to almost all temperate-zone perennial fruit crops. For instance, because of the relatively soft nature and high sugar content of most mature or nearly mature fruits, fruit rots are common afflictions.

Sunlight and circulating air help to dry leaf and fruit surfaces, thereby limiting fungal and bacterial infections.

The organic grower can help to minimize fruit rots by allowing good air circulation and sunlight penetration into the interior plant canopy. In tree crops, this would mean proper pruning and training. In brambles and strawberries, reducing plant density helps. In grapes, adequate pruning and removing leaves that shade fruit clusters is beneficial. All fruit crops need a site that allows good air circulation. Well-timed applications of allowed fungicides can be effective in an integrated disease-control program for mildew and fruit rots in certain fruit crops.

Another problem common to many fruit crops is root rot and intolerance to poorly drained soils. Blackberries, most pear rootstocks, and some apple rootstocks are relatively tolerant of heavy or poorly-drained soils, but even these crops will succumb to persistently water-logged conditions. Blueberries, raspberries, and *Prunus* species (peaches, plums, cherries, etc.) are very intolerant of poorly drained soils and are generally susceptible to root-rotting organisms common in such soils. Even in well-drained soils, blueberries and raspberries are often planted in hills or raised beds. Again, site selection is very important.

Soils can be made disease-suppressive through the addition of significant amounts of organic matter to the soil. This has been most vividly demonstrated in Australia, where liming and cover crops—combined with applications of chicken manure, cereal straw, weed residues, and other materials—are used in avocado groves to control *Phytopthora* root rot. This strategy, known as the "Ashburner system," is now common practice in many areas where avocados are grown. In contrast, mulching apple trees in humid areas, such as New York, may increase *Phytopthora* root rot

To identify fruit diseases and their life cycles, please see fruit production resources compiled by the Cooperative Extension Service. Some of the publications in the Resources section of this publication provide an excellent summary of fruit diseases.

Plant Health and Vigour

Maintaining plants in good health and vigour is important in insect pest management. For fruit plants, this adage is more applicable to indirect pests (those that feed on foliage, stems, etc.) than to pests that feed on the fruit. For instance, an apparently healthy plum tree may set a good crop of fruit, yet lose it all to the plum curculio. That same tree might suffer significant defoliation by caterpillars early in the season; yet, if it is in good vigour, it can compensate and bounce back quickly—still producing a marketable crop that year. There are some cases where general plant health and freedom from stress do impart a form of "resistance"—not technically genetic resistance—to certain pests. Two examples are apple trees in good vigour that actually cast out invading flathead apple tree borers by smothering them with sap, and plants not suffering drought stress being much less attractive to grasshoppers.

Biological Control

Biological control uses living organisms to manage pest populations. When a pest is endemic (not exotic), natural enemies are present, and biological control occurs naturally. The fact that it is occurring may not be noticed by growers. Researchers monitoring certain pests, such as leafminers, have found that pest populations actually increase after pesticide applications kill their natural enemies.

As described above, biological control can be enhanced by cover crops and habitat management. However, where a known pest appears predictably and can be controlled by a specific biological agent, timed releases of beneficial insects may be in order. Many beneficial insects can be purchased from commercial insectaries for release in fruit plantings. Examples of beneficial arthropods that have been used to control pests in fruit crops include the predatory mites *Phytoseiulus persimilis* and *Metaseiulus occidentalis*, which attack spider mites; lady beetles and green lacewings, which feed on aphids; and *Trichogramma* wasps, which parasitize the eggs of several pests, including codling moth.

As a rule, beneficial arthropods are not a complete control for direct fruit pests, at least not for commercial growers who have a low damage threshold for fresh fruit. Usually, additional control measures are necessary. There are four essential components for successful use of beneficial organisms for pest control.

1. *Selection of the proper natural enemy for a target pest.* For example, *Trichogramma* wasps parasitize eggs and, therefore, do not directly control adult pests already active in the field.
2. *Proper timing of releases.* Release of natural enemies must coincide with a susceptible stage of the host and should be made early enough in the cropping season to assure success.
3. *Correct rate of release for natural enemies.* This is usually based on the planting density.
4. *Environmental provisioning.* Make sure environmental needs—such as nectar sources, alternate prey, and water—are available for adult beneficial insects. If the necessary environment is not available, beneficials may leave the release area, die, or spend so much time searching for nectar or pollen that they do not efficiently attack pests.

Applying Materials: Pesticides Allowed in Organic Production

Allowed materials include only natural (non-synthetic) materials that are not specifically prohibited, and specifically allowed synthetic materials, as described in 7 CFR §205.600-602 of the NOP. Most, if not all, allowed synthetic materials have annotations that closely restrict how (in purpose, application method, and quantity) they can be used. Before you apply any product, make sure it's allowed for use in organic agriculture. Read the label carefully. Are all the active ingredients allowed? What about the inert ingredients? If it contains any undisclosed inert ingredients, you must have documentation from the manufacturer to confirm that all inerts are allowed by the National Organic Program (on EPA List 4). If in doubt, ask your certifier before you use it.

Information to guide your selection and use of materials:

- NOP Standards: The National List 205.600-602. This is USDA's ultimate authority. This list, however, can be difficult to use because it is not a positive statement of materials that can and cannot be used. Instead, it states that all natural (non-synthetic) materials are allowed, unless they are prohibited (those materials on list 206.602), and that all synthetic materials are prohibited unless they are allowed, many with restrictions, as named and described on list 205.601. In some instances, it is not clear whether a material is natural or synthetic. For that reason, many organic producers and certifiers also use related resources to help interpret the National Lists and determine the allowability of substances in organic production.
- The Organic Materials Review Institute (OMRI) reviews and evaluates materials for consistency with the National List and, publishes and maintains two other lists. The Generic Materials List is available in hard copy to members for a small fee. These generic and brand-name lists provide an extremely useful resource to help growers identify allowable materials and products. While many organic producers, handlers, and certification agencies rely on and frequently refer to the OMRI lists, users should be aware that OMRI is an independent nonprofit organization, and USDA has not formally recognized or authorized OMRI to interpret the National Organic Standards.

Several new disease-control materials on the market are allowed for use in organic agriculture, including biofungicides, mineral-based essential oil extracts, and botanical fungicides. Growers in some regions are also using compost teas and plant extracts.

The OMRI list provides information about the allowability of brand name products, but not their efficacy. You can ask your local Extension agent about any research or use in your region. Copper and sulfur compounds are fungicides that are allowed (with restrictions on their use) and have been used historically by organic growers, but they have several drawbacks.

These materials can damage plants if applied incorrectly. Sulfur dust can cause acute eye and respiratory irritation in humans. It is also lethal to some beneficial insects, spiders, and mites, and can set the stage for further pest problems. Long-term frequent use of copper fungicides can also lead to toxic levels of copper in the soil.

Fertilization

Fruits, being largely water and sugars, remove relatively few nutrients from the soil, compared to other crops. Therefore, much of a fruit crop's fertility needs can be met through cover crop management and organic mulches (in systems that use them) and by the application of lime and other slow-release rock powders at the pre-plant stage. Supplementary fertilization may still be required for optimal growth and production.

There are many commercial organic fertilizers available. As noted above, however, organic growers should be familiar with organic standards, and especially the National List of Allowed Materials in the NOP's 7 CFR §205.600-602.

The NOP Rule requires that applied raw manure be incorporated at least 90 to 120 days before harvest of crops for human consumption (90 days if the edible portion does not have contact with soil or soil particles; 120 if it does). Compliance is easy: move grazing animals to another pen or paddock at the appropriate time.

General Fertilizer Guidelines

- Organic fertilizers—especially uncomposted animal manures—should be incorporated into the soil to avoid nitrogen volatilization and to comply with organic standards. Use shallow tillage to prevent damage to plant roots and to minimize the potential for soil erosion. Manures should be incorporated into the soil at least three or four months before harvest (depending on the crop type) to comply with National Organic Program standards.
- Soluble organic fertilizers such as fish emulsion, kelp, and soybean derivatives are suitable for use in drip irrigation and can provide quick supplemental fertility. Compost teas may be allowed and may contribute to disease control. Be sure to check with your certifying agency regarding current interpretations of the organic standards for compost production and any restrictions on the preparation or use of compost teas.
- Most organic fertilization programs focus on supplementing nitrogen as the key element, since it is needed in the greatest amount for the crop. You can calculate rates of organic amendments based on standard recommended rates for the crop, but be aware that many fertilizer recommendations still assume the use of synthetic materials. Organic systems behave differently. They generally use slower-release fertilizers and rely on biological activity to break them down into forms that can be absorbed by the plants. For example, only a portion (perhaps half) of the nitrogen applied as stable compost may be available to plants in the first year. The rest is stored and released gradually. To compensate for this, the producer may apply twice as much nitrogen as is needed in the first year of organic management. In subsequent years, however,

more of the nitrogen is released from the soil organic matter and becomes available. In a mature organic farming system, nutrients and organic matter are added to maintain, replenish, and build the bank of nutrients in the soil.

- When making fertilizer calculations based on nitrogen, growers need to credit the estimated contributions made by legume cover crops and/or mulches. A cover crop of subterranean clover, properly fertilized and inoculated, can fix from 100 to 200 pounds of nitrogen per acre annually in a "living mulch" system. Other legume cover crops may produce as much or even more, depending on things such as planting date, weather, and mowing.
- Consider the overall fertilizer analysis; basing application rates solely on nitrogen content can cause problems when the fertilizers are not balanced to meet the needs of the crop. For example, repeated use of poultry manure, which is very high in phosphate, can lead both to pollution problems and to zinc deficiency in the crop. These problems can be avoided by regularly monitoring and adjusting fertilizer selection and rates.
- The most reliable means for determining whether fertilization is adequate is to combine field observations with soil or tissue testing. Poor yields, unusual coloration of leaves, and poor plant growth are all clues to a possible nutritional imbalance or deficiency. On most fruit trees, slow elongation of branches often indicates a nitrogen deficiency. Yellowing between the veins of new blueberry leaves usually means the plant is suffering an iron deficiency. Corky bark on certain apple varieties can indicate an over-availability of manganese in the soil.
- Foliar analysis measures the nutrient content of the leaves and can identify a nutrient deficiency or excess well in advance of visible symptoms. It is more helpful than a soil test because the foliar analysis is a measure of what the plant is actually taking up, while a soil analysis only measures what is in the soil—which may or may not be available to the plant. Annual foliar analysis generally provides the best guide for adjusting supplementary nitrogen fertilization.

Organic apple growers on California's Central Coast say that most years they plant only rye or other grass cover crops, because leguminous cover crops would contribute too much nitrogen, inducing excess growth, creating more pruning work, and decreasing fruit production. Growers monitor their nitrogen levels through leaf-tissue and soil analysis, and manage their soil covers accordingly.

Organic Weed Management

Some weed control methods, such as smother crops, are discussed in the Site Preparation section above. This type of cover cropping is an important tool for weed management that also contributes to good soil management, fertility, and pest management.

Mulches

Organic Mulch

Mulching is a powerful weed management strategy that can also contribute to good soil management, if appropriate natural materials are used. After a planting is established, weeds can be suppressed by applying thick layers of mulch. This can also create habitats for beneficial arthopods, including generalist predators such as big-eyed bugs, soft-bodied flower beetles, and spiders. Organic mulches are usually applied in a circle around tree trunks or vines, and down the whole row in blueberries.

Commonly, tree fruit growers keep mulches away from the tree trunks, particularly in winter, to prevent voles or mice from gnawing on the bark and damaging young trees. Keeping mulches 8 to 12 inches away from the trunk also reduces the likelihood of crown rot and other diseases in susceptible species—most notably apples on certain rootstocks.

Mulch materials may include straw, spoiled hay, leaves, yard trimmings, woodchips, and sawdust. Many of these materials are inexpensive. Still, it's wise to weigh the benefits and risks of each, including hauling costs and the risks of their containing impurities and prohibited materials.

Municipal greenwaste may be available, either raw or from municipal or commercial composting operations. Growers must monitor the incoming product and remove any trash to keep undesirable material out of their fields. Growers should ask compost producers about the sources of their materials and any pesticides that may persist in them. Of particular concern are clopyralid and picloram, herbicides that are extremely resistant to breakdown, even after composting. The sale and use of these materials is restricted in some areas. A Washington State University study showed treated grass clippings to be the primary source of clopyralid entering the organic waste stream. Experience from California, Oregon, and Washington shows that at levels of 1 to 10 parts per billion, clopyralid adversely affects sensitive vegetable crops.

Because organic mulches decompose over time, they require periodic re-applications in order to continue suppressing weeds. However, their decomposition provides other benefits. Mulching with organic matter enhances soil aggregation and water-holding capacity. Researchers from 1937 to the present have consistently found that mulching is the best orchard-floor management system for retaining moisture.In Michigan research, mulching was as effective as irrigation in encouraging tree growth.Organic mulches can have positive effects on tree growth, with improvements in soil quality and shifts toward beneficial nematodes. Mulch can also benefit the crop by moderating soil temperatures, thus reducing plant stress. Organic mulches provide slow-release nutrients for the long-term health and fertility of the soil. Research indicates that potassium, phosphorus, and nitrogen (primarily from the slow breakdown of the mulch) are more available in mulched systems than in non-mulched systems.Some growers express concern that sawdust may acidify their soil or bind nitrogen in the soil. However, these effects are minimal if the sawdust is not tilled into the soil. Raising organic matter on the farm is one way to ensure sufficient, clean mulching material. Farm-raised hay grown outside the orchard can provide weed-free mulch. Cover crops may be grown between tree rows, mowed, and gathered around the trees. Some small-scale growers use the biomass from orchard alleyways, cutting cover crops with a sickle-bar mower and hand-raking the material under the trees. Larger-scale operations often use forage wagons, straw-bale spreaders, or specialized equipment to mechanize mulching jobs. King Machine Co. offers a small, trailer-or truck-mounted square-bale chopper and blower suitable for most fruit crops. Millcreek Manufacturing Co. has developed a row mulcher especially suitable to blueberry, bramble, and grape culture, but also useful in tree fruit orchards. The Millcreek machines are designed to handle bulk organic materials such as sawdust, wood chips, bark, peat, and compost.

Geotextiles

Geotextile mulches are paper or woven plastic fabrics that suppress weed growth. While they allow some air and water penetration, they may reduce water infiltration, whereas organic mulches increase infiltration.Geotextile mulches do not provide the advantages of adding matter and nutrients to the soil, and if synthetic, they must eventually be removed. Geotextiles have a high intial cost, though this may be partially recouped in lower weed control costs over the materials expected field-life—5 to 10 years for polyester fabric; 2 to 3 years for paper weed barriers. Still, some growers find them useful for weed suppression in orchard, tree plantations, and cane fruit culture.

Sheet Mulch

You can also create weed barriers by sheet mulching: laying down layers of cardboard or newspaper and covering them with organic material. Sheet mulching increases the efficacy of organic mulch as a barrier against emerging weeds. Organic growers should avoid cardboard that is waxed or impregnated with fungicide, as well as colour print and glossy paper, in order to be compliant with the National Organic Program standards.

Cultivation

Cultivation—using mechanical tillage and weed harrowing implements—is the most widely used weed-management practice in fruit production. In systems that maintain permanent vegetation between rows, cultivation may be limited to the tree row under the dripline in an orchard, or extended 1 to 3 feet from the edge of the hedgerow in bramble plantings. The reverse is true where mulches are used in the tree row, and cultivation is used to control weeds and incorporate cover crops in the alleyways. In any case, cultivation must be kept shallow to minimize damage to crop roots and to avoid bringing weed seeds to the surface.

Hand cultivation—enhanced with the use of a wheel hoe—can be effective in small-scale plantings. In large-scale plantings of trees or vines, where in-row tillage is desired, "mechanical hoes" such as the Weed Badger or Green Hoe are very useful. These tractor-mounted, PTO-driven cultivators can till right up to the tree or vine without damaging the plant. Attachment options include powered rotary tillage tools and scraper blades that can move soil either away from or toward the base of the crop plants. Scraper-blade attachments, commonly known as "grape hoes," have been used in vineyards for decades.

Herbicides Allowed for Use in Organic Production

A few herbicides currently emerging on the market are allowable for organic production, with restrictions on the location of their use. There is ongoing research on using materials such as vinegar, corn gluten, and citric acid as herbicides, although they are not yet widely used by certified organic growers. Such materials may have applications in organic systems, such as for spot treatment of noxious weeds.

Weeder Geese, Chickens, and Ducks

For many years, farmers have used geese to control weeds in perennial and annual crops, including strawberries, blueberries, bramble fruits, and tree orchards. In Oklahoma, researchers at the Kerr Centre for Sustainable Agriculture used weeder geese in commercial-scale blueberry and strawberry production, with portable electric fencing to keep the birds in a specific zone in the plant row.Investigators at Michigan State University studied the impacts of populations of domestic geese and chickens in a nonchemical orchard system. They found that the geese fed heavily on weeds—especially grasses—and also on windfall fruit. In general, geese are more effective against emerging or small grass weeds, and they have a particular preference for Bermuda grass and Johnson grass—weeds that can be especially troublesome in orchards.

Those who have raised chickens know how enthusiastically they devour fresh vegetation. If the area they inhabit is small, they will strip it to the dirt. Properly managed, however, their foraging characteristics can be used to the grower's advantage.

Fred Reid is an innovative producer of raspberries and vegetables in Canada who has successfully employed his flocks of chickens in weed management. He uses a system of fencing to keep chickens in certain areas to accomplish a thorough job of weeding and insect control. He notes that if the vegetation has grown too high and the plants become too fibrous, the chickens will not eat them. However, if you mow tall vegetation in advance, the chickens will process it readily. He excludes the chickens from raspberry plots when the new, tender leaves are emerging and, of course, near harvest time.

Flame Weeding

Flame cultivation uses directed heat to kill weeds. It works not by burning the weeds but by searing them and causing the plant cells to rupture. Farmers began using tractor-mounted flamers in orchard and row crops in the 1940s. Technology and technique have both been refined considerably in recent years. Several tools now commercially available, including flame, infrared, and steam weeders, make heat a viable option for some weed management applications.

Management of Vertebrate Pests: Mammals and Birds

Several bird species, deer, rabbits, ground squirrels, gophers, mice, voles, raccoons, and other animals can be serious pests of fruit plantings. Organic certification calls for an integrated approach to vertebrate management, including exclusion, trapping, repellents, scare devices, and protection or development of predator habitat.

Gophers and ground squirrels can be managed on organic farms through integrated strategies. Thomas Wittman of Gophers Limited emphasizes that growers should not expect to eliminate these pests, but will do well to keep populations in check. He stresses the importance of keen observation and has tips for effective trapping routines using commercially available traps. Persistent year-round trapping is the primary strategy for most farmers, complemented by enhancing the habitat of key predators such as owls and hawks with nestboxes, perches, and appropriate vegetation.

Explosive propane devices are effective against gophers and ground squirrels. Propane gas ignited in rodent burrows creates an explosion that kills the animals and disrupts their tunnels. Several organic orchardists say that this works, but most promptly abandoned its use because neighbors complained about the noise of the explosion, similar to the sound of a gun shot. Only two materials (sulfur dioxide for underground smoke bombs and Vitamin D3, or Cholecalciferol) are on the National List as rodenticides. These may be used only if they are documented in the Organic System Plan, used with care to avoid harming non-target animals, and only when other management practices are ineffective.

Birds can be especially troublesome in cherry, berry, and grape plantings. Exclusion with bird netting is probably the surest control, but the initial cost can be high for both the material and its placement. Noise devices, "scare-eye" balloons, Mylar tape, artificial hawk kites, and many other home remedies have been tried, with varied success. Successful scare tactics depend on the bird species, bird population pressure, and the grower's management of the devices. It is important to remember that birds and other vertebrate pests are quick to learn, and they often overcome their initial aversion to scare devices or repellents. The grower will achieve the most effective control by moving devices frequently, and by changing or mixing the devices. For example, organic growers describe how they effectively scare certain bird species away from newly emerging crops by placing red- and silver-coloured Mylar tape in the field for just a few weeks, then removing it so that the birds do not get used to seeing it. Fruit growers use sonic and visual scare devices only at critical times in the growing season, such as fruit ripening, and remove them promptly as soon as that period is over.

Deer can be devastating to fruit plants, especially young orchards. Methods for preventing or controlling deer damage to crops range from exclusion and cultural methods to scare devices, repellents, and culling or harvest.

Deer Control Options. Scent and taste repellents may be effective under light deer pressure, especially if the grower switches periodically from one repellent or deterrent to another. Research at the University of Wisconsin indicates that none of these repellents is very useful under heavy deer pressure. Exclusion fencing may be the only way to manage heavy deer populations. In most states, the Cooperative Extension Service provides suitable plans for deer fencing. Electric fencing appears most effective. Research indicates that even a single strand of electrified wire can work. Where deer problems are severe, however, a seven-strand, sloped, electrified fence may be necessary.

Tree guards made of plastic, hardware cloth, or similar materials can keep rabbits from gnawing on fruit tree trunks.

However, northern growers should remember that snow can effectively raise the gnawing height of rabbits.

Mice and voles may be attracted to mulch around fruit plants. Such rodents take up residence in mulch during the winter, feeding and gnawing on roots, stems, and trunks. To reduce the chance of vole damage, mulch should be raked away from the plants in the fall (usually 18 to 24 inches is adequate). Mulch removal may not be practical, however, for blueberry plantings. Keeping the planting site mowed also helps reduce rodents by exposing them to natural predators such as hawks and owls. For pests such as raccoons, opossums, skunks, etc., tight web-type fencing or non-lethal traps are the best control options.

Postharvest Handling

Many fruits require some type of postharvest handling. Whether done on-farm or off, these processes must be documented in the Organic System Plan. Any off-farm postharvest handling must be done by certified organic facilities, and appropriate measures must be taken to prevent commingling or contamination of organic products with non-organic products during washing, sizing, packing, and storage. A complete audit trail must track produce from its field of origin to the point of final sale. Growers should read the National Organic Program Final Rule, review the National List, OMRI lists, and consult with their certifiers about any materials to be used postharvest, such as cleansers, shellacs, or waxes.

Conclusion

Organic fruit production is a practical option for some growers, but the viability of the enterprise will likely hinge on site, scale, type of fruit, markets, and managerial skills. In general, crops grown in the drier climates of the West have fewer disease and pest problems. This region may, therefore, be better suited to organic fruit production. Strawberries, bush fruits, and brambles are probably easier to grow organically in most sections of the country than grapes and tree fruits. Management requirements for organic production are likely to be higher in any region, and the producer must be closely attuned to local site conditions.

Bibliography

Arguello D. S.: *Production, Post-harvest Management and Exportation of Tropical Fruits in Costa Rica,* Wageningen, Technical Centre for Agricultural and Rural Cooperation, 1992.

Ashraf, Syed Mahboob: *Handbook of Fruit and Vegetable Products,* Agrobios, Delhi, 2008.

Ashworth S.: *Seed to Seed,* Decorah, Seed Savers Publications, 1991.

Barbeau G.: *Tropical Fruits in Nicaragua,* Managua, Nicaragua Ministerio de Desarrollo Agropecuario, Agraria, 1990.

Berger, P.L.: *Pyramids of Sacrifice: Political Ethics and Social Change,* New York, Basic Books, 1974.

Bhutani, R.C.: *Fruit and Vegetable Preservation,* Biotech Books, Delhi, 2003.

Cao Van P.: *An Integrated Approach for the Production and Processing of Minor Fruits in Martinique (FWI),* Port of Spain, Trinidad and Tobago, 1996.

Chadha K. L. and Pareek O. P.: *Advances in Horticulture: Fruit Crops,* New Delhi, Malhotra Publishing House, 1993.

Collymore L.: *Fruit Production in Barbados,* Port of Spain, Trinidad and Tobago, 1996.

Crucefix D.: *Avocado Variety Selection for Export Development,* Roseau, CARDI, 1996.

Cull Brian & Lindsay Pax: *Fruit Growing in Warm Climates for Commercial Growers & Home Gardeners,* Australia, Reed Books, 1995.

Currah L. and Proctor F. J.: *Onions in Tropical Regions,* Kent, Natural Resources Institute, 1990.

Degras L.: *Yam: a Tropical Root Crop,* Wageningen, CTA/MacMillan, 1993.

Dhatt A. S. and Singh Z.: *Propagation and Rootstocks of Citrus,* New Delhi, Malhotra Publishers, 1993.

Doijode S. D.: *Seed Germination in Fruits,* New Delhi, Malhotra Publishers, 1993.

Fageria, N K ; V C Baligar and R B Clark: *Physiology of Crop Production,* International Book, Delhi, 2007.

Harwood, R. R.: *Marshalling Technology for Development: Proceedings of a Symposium,* Washington, DC, National Academy Press, 1995.

Herminie Broedel Kitchen: *Soils and Crops : Diagnostic Techniques,* Satish Serial Publishing, Allahabad, 2004.

Hessayon D. G. Dr.: *The Vegetable Expert*, England, PBI, Publications, 1985.

Hurst Jacqui & Rutherford Lyn.: *A Gourmet's Book of Mushrooms & Truffles*, Sydney, Golden Press Pvt. Ltd., 1991.

Julia F.: *Fruits of Warm Climates*, Miami, Julia F. Morton Publisher, 1987.

Mabberley D. J.: *The Plant-Book : a Portable Dictionary of the Vascular Plants*, Cambridge, Cambridge University Press, 1997.

Malins A.: *Postharvest Handling of Pineapple and Mango*, Port of Spain, Trinidad and Tobago, 1992.

Matsuoka H.: *Cultivation of Panicum Genetic Resources for Evaluation of Characteristics*, Tokyo, JICA, 1997.

Mitra S.: *Postharvest Physiology and Storage of Tropical and Subtropical Fruits*, Oxon, CABI, 1997.

Murali T. P. and Duncan E. J.: *In Vitro Propagation of Banana through Aseptic Manipulation of Male Inflorescences*, Port of Spain, NIHERST, 1990.

Oldham P.: *Cost of Production of Major Tree Crops in Dominica*, Roseau, Ministry of Agriculture, 1991.

Pilgrim R.: *Post Harvest Handling of Minor Exotics*, St. George's, Grenada, 1996.

Rehm Sigmund: *Multilingual Dictionary of Agronomic Plants*, Boston, Kluwer Academic Publishers, 1994.

Sinha G. C., Reddy Y. T. N. and Singh G.: *Propagation and Rootstocks in Guava*, New Delhi, Malhotra Publishers, 1993.

Somasundaram Rajarathnam: *Advances in Preservation and Processing Technologies of Fruits and Vegetables*, New India Publishing Agency, Delhi, 2011.

Sreekumar V., Indrasenan G and Mammen G.: *Studies of the Quantitative and Qualitative Attributes of Ginger Cultivars*, Calicut, Kasaragod, 1990.

Srivastava, A.P.: *Mechanization of Vegetable Production and Post-Harvest Management*, Westville Pub, Delhi, 2009.

Thomas E.: *Fruit Production in St. Kitts and Nevis*, Port of Spain, IICA, 1996.

Tripathi, Pramila: *Botanical Pesticides in the Management of Post-harvest Fruit Diseases*, Daya, Delhi, 2005.

Verma, L.R. and Sharma, R.C.: *Diseases of Horticultural Crops: Fruits*, Indus, Delhi, 1999.

Webb M.: *Weed Problems and their Control in Rice, Papaya, Citrus and Sugar cane: a Report to the Belize Plant Protection Service*, United Kingdom, Natural Resources Institute, 1993.

Whitwell A.: *Dominica Orchard Crop Management and Research Project: Pest and Disease Management*, United Kingdom, Natural Resources Institute, 1991.

Yoshida T.: *Cultivation of Citrus Genetic Resources for Evaluation of Characteristics*, Tokyo, JICA, 1996.

Index

□□□